Skills: • identifies word names for numerals
• identifies place value to tens place
• counts by 2's, 3's, and 4's to 100

Examples: the word name for 5847 is five thousand eight hundred forty-seven

Directions: Select or write the correct answer for each question.

1. Which number is five thousand twenty-three?
a. 5,230 b. 5,023 c. 5,203 d. 5,123

2. Write the word name for 9,406.

____

3. Circle the digit in the tens place in each number.
3940 837 7491 95

4. Write the numbers for counting by 2 's from 60 to 100.

____

____

____

5. Write the numbers for counting by 3 's to 51.

____

____

____

6. Write the numbers for counting by 4 's to 60.

____

____

____

REVIEW

Skills: • orders and compares value of numbers
• writes ordinals to the 20th
• identifies odd and even numbers through 99
• recognizes and continues number patterns

Directions: Select or write the correct answer for each question.

1. Rewrite the numbers in order from greatest to smallest:

4893 849 1928 924 21,486 395

Greatest Smallest

_____ _____ _____ _____ _____ _____

2. Use the digits 8, 5 and 2 to make six different numbers. Write them in order from greatest to smallest (one is done for you).

Greatest Smallest

_____ _____ _____ _____ _____ 258

4. Circle the odd numbers, underline the even numbers:

335 92 391 1584 479 54

386 938 396 43 8

5. Continue the patterns:

seventy-one seventy-three seventy-five __________ __________

1397 1396 1395 1394 1393 ________ ________ ________ ________

5 8 11 14 17 ________ ________

Tenth Eleventh Twelfth ______________ ______________

REVIEW

Skills: • uses mathematical symbols correctly
• identifies and writes Roman numeral values
• rounds numbers to the nearest ten or nearest hundred

Examples: 76 rounded to the nearest ten is 80
931 rounded to the nearest ten is 930

Directions: Select or write the correct answers.

1. Complete the number sentences using > < = - + x ÷:

79 ◯ 22　　103 ◯ 191　　4 ◯ 9 = 13　　72 ◯ 2 = 70

3 ◯ 9 = 27　　9 ◯ 3 = 3　　4 + 4 - 1 ◯ 7

2. Write the Roman numerals for each number:

four ______　　31 ______

19 ______　　eight ______

3. Round the numbers to the nearest ten:

349 ______　　21 ______　　58 ______

4. Round the numbers to the nearest hundred:

733 ______　　289 ______　　955 ______

Skills: • estimates by rounding to the nearest ten
• recognizes and writes numbers in expanded form

Example: expanded form: 467 = 400 + 60 + 7

Directions: Estimate and use mental math to solve the problems.
Do not use a pencil or a calculator for the first two questions.

1. The students were collecting canned food for flood victims. Sara brought 9, Matt brought 12, Lena brought 8, and Colleen brought 10. Without using a calculator or paper and pencil, estimate the total number of cans they brought.

a. about 40 cans   b. about 30 cans   c. about 50 cans   d. about 20 cans

2. In the Walk-a-Thon, 99 women walked, 81 men strolled, 102 boys and 101 girls ambled. About how many people completed the Walk-a-thon?

____

Directions: Rewrite each number in expanded form.

3. 869 - ____________________

5372 - ____________________

Directions: Read the expanded form; then write the number.

4. 900 + 40 + 8 ____________________

6000 + 400 + 70 + 2 ____________________

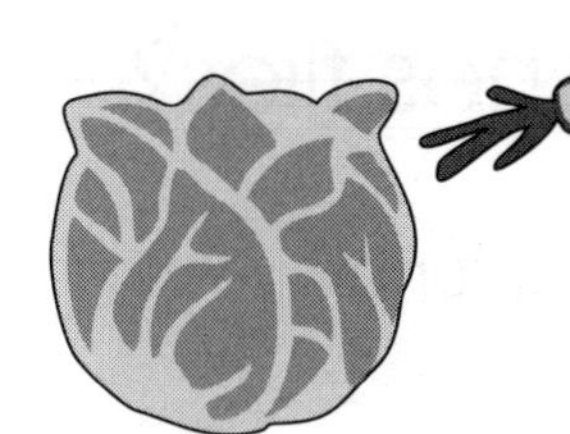
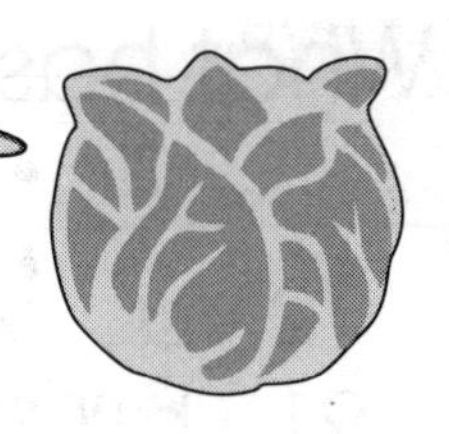

Add these numbers.

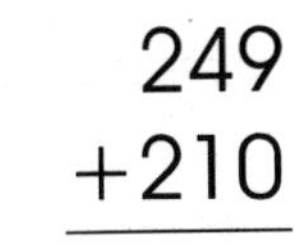
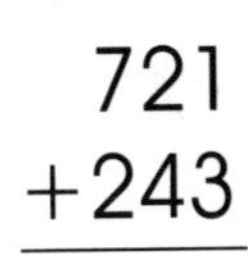

249 +210	582 +312	213 +754
721 +243	323 +573	620 +145
143 +346	523 +316	251 +120
354 +621	202 +104	313 +514
436 +122	510 +374	421 +436

What has 18 legs, red spots and catches flies?

72 - a	99 - h	64 - s
91 - b	93 - i	37 - t
83 - e	45 - L	56 - w
	58 - m	

49 + 23
72
a

74 + 17	27 + 45	37 + 27	55 + 28	35 + 56	56 + 16	27 + 18	19 + 26
91							
b							

29 + 8	37 + 46	35 + 37	19 + 39

28 + 28	64 + 29	19 + 18	59 + 40

26 + 32	29 + 54	60 + 12	46 + 18	15 + 30	66 + 17	15 + 49

Check Your Answers

$\begin{array}{r} {\scriptstyle 11} \\ 364 \\ +278 \\ \hline 642 \end{array}$ $\quad$ $\begin{array}{r} {\scriptstyle 11} \\ 278 \\ +364 \\ \hline 642 \end{array}$	$\begin{array}{r} 158 \\ +465 \\ \hline \end{array}$ $\quad$ $\begin{array}{r} 465 \\ +158 \\ \hline \end{array}$
$\begin{array}{r} 297 \\ +627 \\ \hline \end{array}$	$\begin{array}{r} 544 \\ +359 \\ \hline \end{array}$
$\begin{array}{r} 472 \\ +359 \\ \hline \end{array}$	$\begin{array}{r} 299 \\ +595 \\ \hline \end{array}$
$\begin{array}{r} 415 \\ +398 \\ \hline \end{array}$	$\begin{array}{r} 268 \\ +298 \\ \hline \end{array}$
$\begin{array}{r} 273 \\ +189 \\ \hline \end{array}$	$\begin{array}{r} 378 \\ +443 \\ \hline \end{array}$

The seeds in my bird food are very small. Add to find how many seeds are in each bag below.

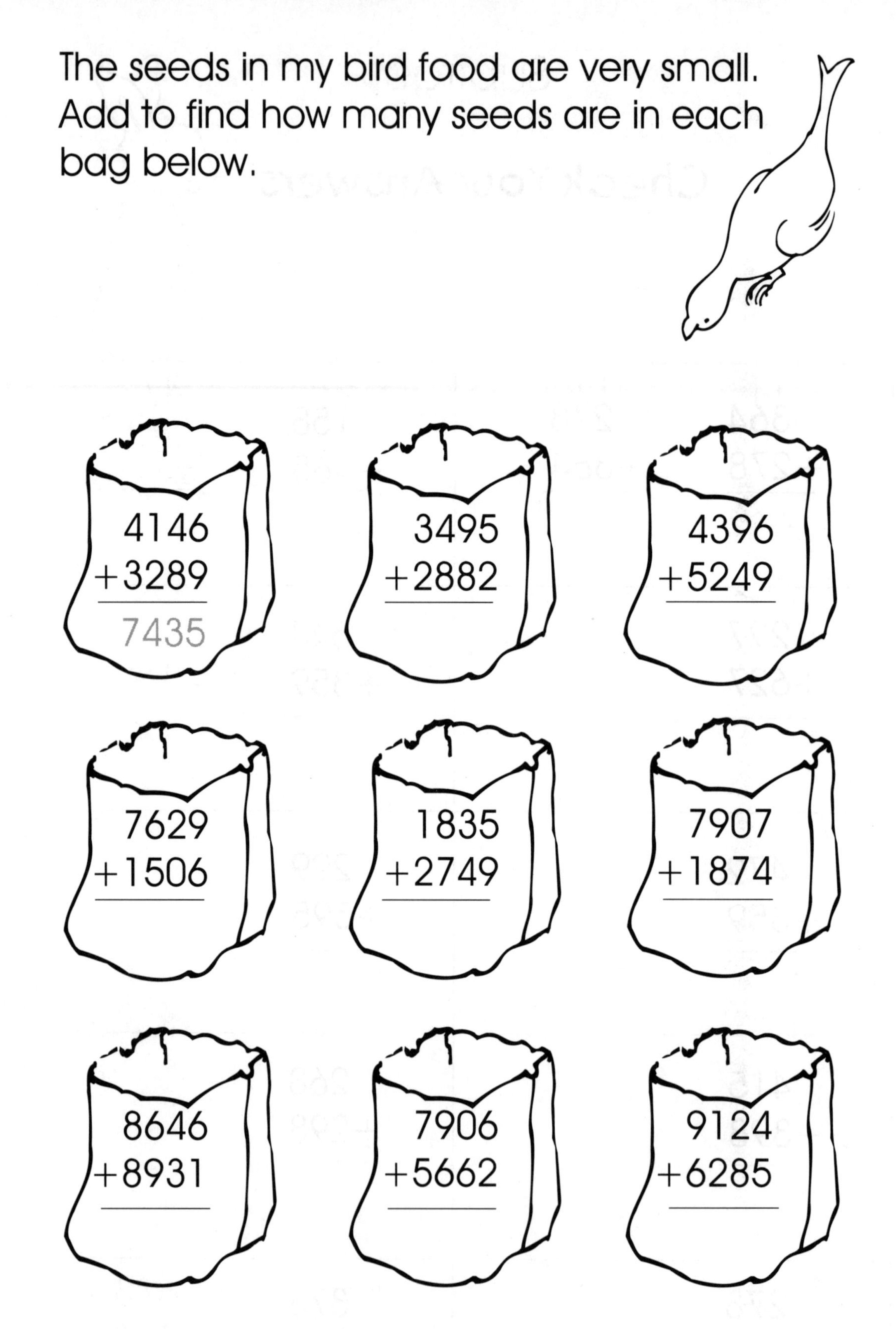

Put an X on the bag with the most seeds.
Put a circle around the bag with the fewest seeds.

Subtract

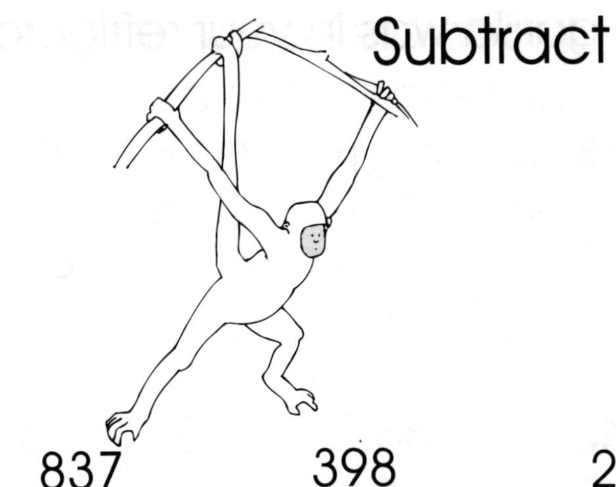

837 - 614 223	398 - 235	254 - 122	473 - 351
256 - 142	638 - 231	356 - 214	189 - 123
643 - 331	532 - 210	975 - 134	837 - 213
584 - 121	324 - 102	949 - 325	613 - 312

How do you know a gorilla was in your refrigerator?

18 - b	44 - h	78 - o	89 - s
56 - e	15 - i	63 - p	29 - t
36 - f	47 - n	27 - r	62 - u

3 13 ~~43~~ - 7	86 - 8	93 - 15	36 - 7
36			
f			

71 - 8	66 - 39	24 - 9	56 - 9	53 - 24	98 - 9

52 - 37	91 - 44

45 - 16	52 - 8	63 - 7

45 - 27	81 - 19	37 - 8	54 - 25	72 - 16	43 - 16

Help this little monkey answer some riddles.

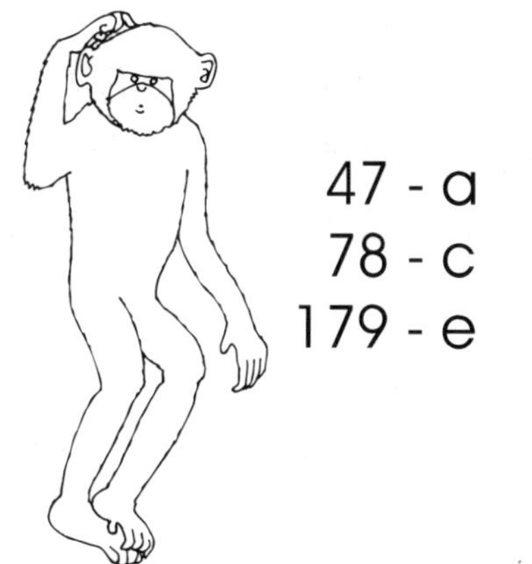

47 - a	227 - h	395 - r
78 - c	245 - n	427 - s
179 - e	338 - o	577 - t
	387 - p	

1. What has eyes but cannot see?

5 12 ~~662~~ - 275	504 - 166	734 - 157	426 - 379	845 - 268	615 - 277
387					
p					

2. What has ears but cannot hear?

356 - 278	723 - 385	783 - 388	912 - 667

3. What has a tongue but cannot talk?

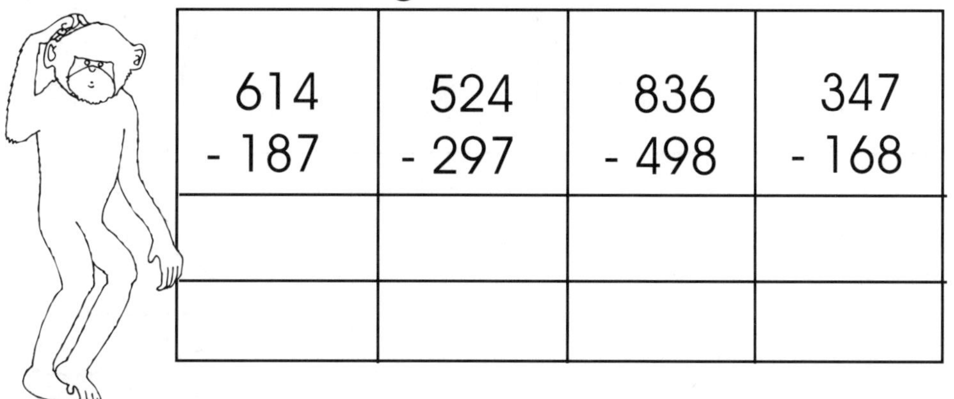

614 - 187	524 - 297	836 - 498	347 - 168

Check Subtraction with Addition

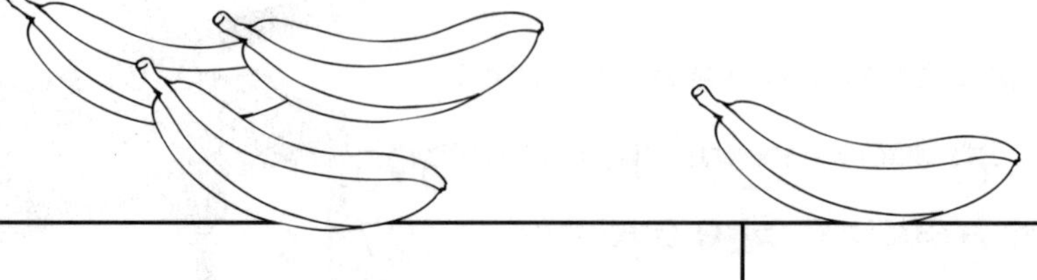

3 16 2 12 ~~4632~~ - 1708 2924 1 1 2924 + 1708 4632	5215 - 2166
9478 - 2294	7685 - 2827
7605 - 4138	9253 - 3618

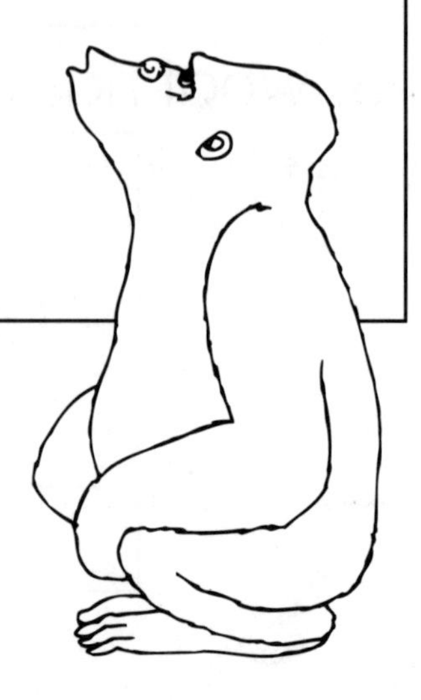

In Search of Treasure

Many years ago, pirates buried treasure on a small island in the ocean. Captain Jones found an old map showing where to find the treasure. She and her crew decided to look for the treasure. Find the answers to these problems to help them.

Problem	Work Area
The crew landed on shore. They walked 6 miles to a large boulder. They turned left and walked 5 miles to some palm trees. They rested for the night. The next day they walked 8 more miles. How far did they walk?	
The crew came to a small lagoon. There was a small hill in the middle of the lagoon. It took 2 minutes to row to the hill. It took 6 minutes to climb the hill. It took 43 minutes to dig up the treasure. How many minutes did all of this take?	
When Captain Jones opened the treasure chest, she found 12 bars of gold, 8 long silver chains, and 4 diamond necklaces. The captain got half of the treasure. How much did she get to keep? _____ bars of gold _____ silver chains _____ diamond necklaces	

The Solar System

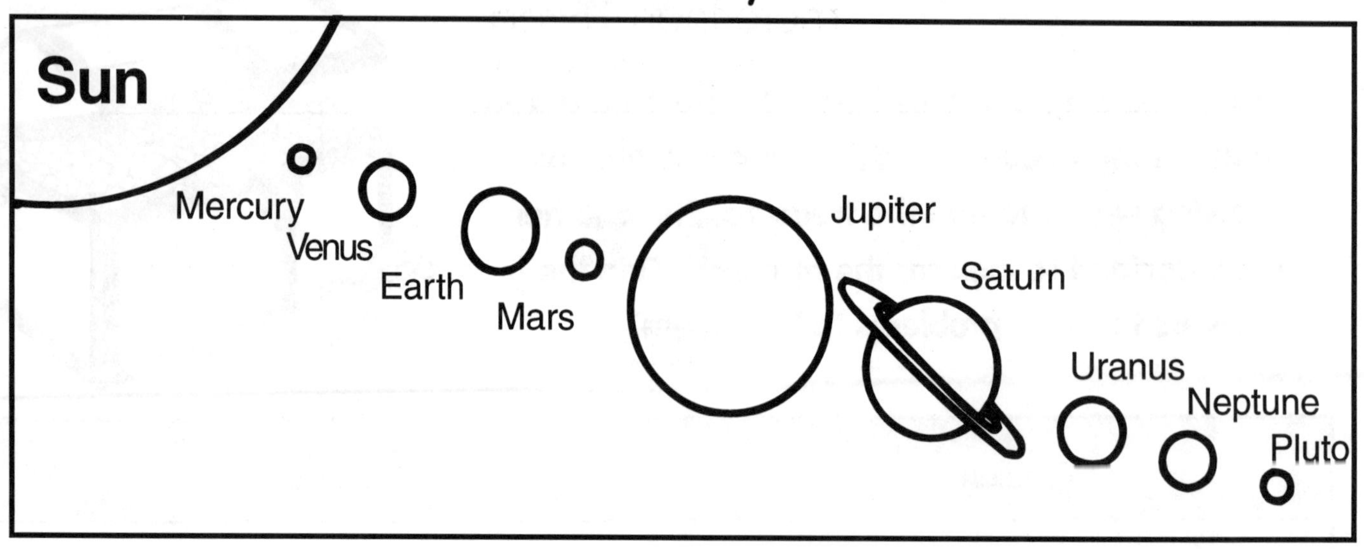

Problem	Work Area
Mercury is 58 million kilometers from the sun. Earth is 155 million kilometers from the sun. How much farther away from the sun is Earth than Mercury?	
The diameter of Earth is 12,756 kilometers. The diameter of Saturn is 120,600 kilometers. The diameter of Uranus is 51,300 kilometers. Is the sum of these three planets more or less than Jupiter's diameter of 142,200 kilometers? How much more or less?	
Pluto is 5,900 million kilometers from the sun. How many kilometers would a space ship have to travel to go from Pluto to the sun and back?	

Color the spaces with answers:
more than 15 red
less than 15 black

6 X 5

9 X 4

2 X 5 = ☐

4 X 4

6 X 3

6 X 5

8 X 3

1 X 3

4 X 3

7 X 5

6 X 2

9 X 2

6 X 4

2 X 5

2 X 4 = ☐

9 X 4

2 X 5

8 X 4

9 X 3

7 X 4

4 X 3

6 X 3

7 X 2

5 X 4

9 X 5

5 X 2

8 X 3

3 X 2

8 X 5

9 X 1

3 X 3

6 X 4

3 X 4

1 X 4

Multiply.

7×8	6×5	4×9	9×6	4×8	9×7
5×7	5×8	6×6	9×9	4×7	4×6
8×6	5×9	8×9	7×7	7×6	8×8

8 X 9 =	5 X 7 =	9 X 9 =
6 X 7 =	4 X 9 =	8 X 6 =
5 X 9 =	8 X 8 =	7 X 7 =
6 X 6 =	7 X 7 =	5 X 6 =
4 X 7 =	5 X 8 =	9 X 7 =

Two-Digit by One-Digit

Here's how to multiply a two-digit number by a one-digit number with regrouping (carrying):

1.
```
 ①13
  X 4
    2
```
Multiply the ones, 3 X 4 = 12
Write the 2 in the ones place.
Write the 1 ten so you remember it.

2.
```
 ①13
  X 4
   52
```
Multiply the tens by 4
1 ten X 4 = 4 tens
Add the 1 ten = 5 tens

13	12	14	19
X 4	X 8	X 7	X 5
52			

18	16	14	17
X 9	X 6	X 3	X 9

24	23	24	30
X 2	X 3	X 2	X 5

42	31	24	62
X 4	X 7	X 4	X 2

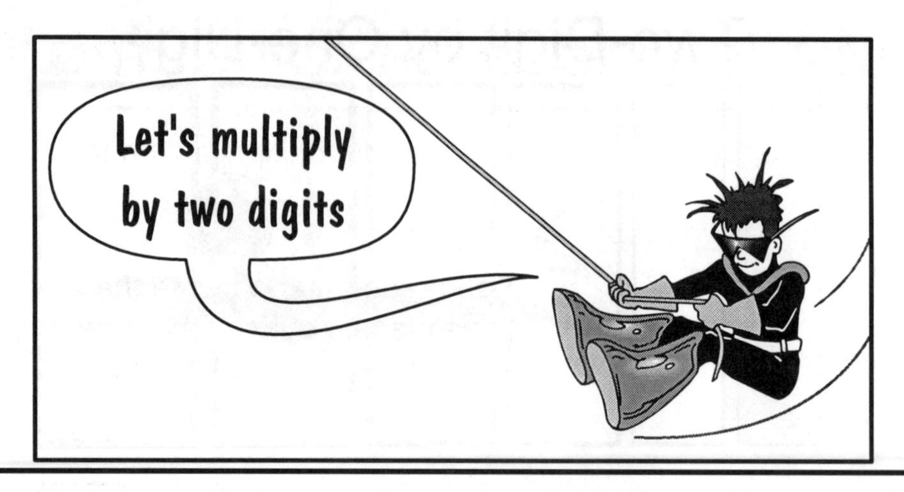

Here's how to multiply a two-digit number by a two-digit number:
Our example problem is:

1.
```
  12        12
 X23       X23
 ---       ---
   6         6
            30
```

2.
```
  12        12
 X23       X23
 ---       ---
   6         6
  30        30
  40        40
           200   Careful - this is 10 X 20 = 200
           ---
           276   Add to get the answer
```

Here's a shorter way:

1.
```
  12
 X23
 ---
  36  (12 X 3)
```

2.
```
  12
 X23
 ---
  36
 240  (12 X 2 tens)
```

3.
```
  12
 X23
 ---
  36
 240
 ---
 276  Add
```

```
  12        41        32        24
 X23       X12       X28       X21
 ---       ---       ---       ---
  36
 24
 ---
 276
```

```
  30        42        40        32
 X32       X21       X12       X14
 ---       ---       ---       ---
```

```
  33        24        32        10
 X12       X32       X24       X10
 ---       ---       ---       ---
```

Multiplying Two-Digit Numbers

Here's a shortcut you can use when you multiply by the tens:

Look at these two problems:

```
  54        54
X 11      X 11
----      ----
  54        54
 540       54□
----      ----
 594       594
```

You do not need to write the zero in the ones places, but you must remember to leave the ones place emply.

```
   3
  34
X 29
----
 306
 68
----
 986
```

34 × 29	23 × 17	83 × 33	92 × 72
45 × 18	72 × 92	64 × 18	50 × 47
52 × 25	51 × 36	90 × 37	35 × 28

$9 \div 3 =$ ___ $2 \div 2 =$ ___ $8 \div 4 =$ ___

$45 \div 5 =$ ___ $24 \div 3 =$ ___ $25 \div 5 =$ ___

$18 \div 6 =$ ___ $56 \div 7 =$ ___ $48 \div 8 =$ ___

$27 \div 3 =$ ___ $64 \div 8 =$ ___ $9 \div 3 =$ ___

$21 \div 7 =$ ___ $18 \div 3 =$ ___ $24 \div 8 =$ ___

$12 \div 4 =$ ___ $16 \div 8 =$ ___ $54 \div 6 =$ ___

$16 \div 8 =$ ___ $40 \div 5 =$ ___ $20 \div 4 =$ ___

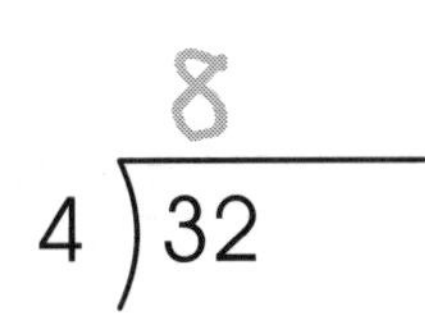

$4\overline{)32}$ = 8	$5\overline{)15}$ = 5	$3\overline{)27}$ = 9
$6\overline{)12}$ = 2	$2\overline{)18}$ = 8	$9\overline{)63}$ = 5
$8\overline{)48}$ = 8	$1\overline{)9}$ = 9	$7\overline{)49}$ = 7
$9\overline{)45}$ = 8	$3\overline{)12}$ = 3	$8\overline{)72}$ = 9
$2\overline{)12}$ = 6	$7\overline{)28}$ = 4	$9\overline{)54}$ = 8
$8\overline{)48}$ = 8	$9\overline{)18}$ = 3	$8\overline{)40}$ = 4

$3\overline{)66}$	$2\overline{)86}$	$9\overline{)90}$	$2\overline{)26}$
$2\overline{)46}$	$4\overline{)84}$	$2\overline{)24}$	$3\overline{)93}$
$7\overline{)77}$	$3\overline{)63}$	$2\overline{)64}$	$9\overline{)99}$
$4\overline{)88}$	$8\overline{)88}$	$3\overline{)36}$	$2\overline{)64}$
$6\overline{)60}$	$2\overline{)42}$	$4\overline{)48}$	$8\overline{)80}$

$3\overline{)69}$

Two-Step Divison

$$\begin{array}{r} 25 \\ 3\overline{)75} \\ \underline{6} \\ 15 \\ \underline{15} \end{array}$$

$6\overline{)78}$

$2\overline{)54}$

$6\overline{)90}$

$3\overline{)78}$

$5\overline{)70}$

$5\overline{)85}$

$6\overline{)72}$

$4\overline{)72}$

$7\overline{)91}$

$4\overline{)96}$

$3\overline{)84}$

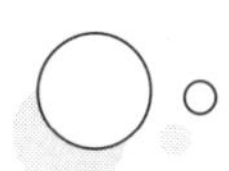

A Sea Riddle

How can you tell a shark has been swimming in your bathtub?

518 - a	921 - k	908 - p
403 - e	510 - m	906 - r
905 - h	275 - n	735 - s
311 - i	487 - o	211 - t

211 $5\overline{)1055}$	$4\overline{)1612}$	$6\overline{)2418}$	$8\overline{)1688}$	$3\overline{)2715}$
t				

$7\overline{)3570}$	$5\overline{)2590}$	$5\overline{)4530}$	$3\overline{)2763}$	$6\overline{)4410}$

$5\overline{)1555}$	$8\overline{)2200}$

$7\overline{)1477}$	$6\overline{)5430}$	$9\overline{)3627}$

$9\overline{)6615}$	$6\overline{)2922}$	$4\overline{)2072}$	$9\overline{)8172}$

Show the Remainders

$$\begin{array}{r} 342\text{ R}1 \\ 2\overline{)685} \\ \underline{6} \\ 8 \\ \underline{8} \\ 5 \\ \underline{4} \\ 1 \end{array}$$

$5\overline{)549}$

$7\overline{)709}$

$8\overline{)489}$

$3\overline{)695}$

$6\overline{)364}$

$3\overline{)908}$

$9\overline{)813}$

$4\overline{)289}$

$7\overline{)744}$

$2\overline{)483}$

$4\overline{)559}$

Check Division with Multiplication

$5\overline{)110}$ — 22; 10; 10; 10 — Check: 22 X 5 = 110	$7\overline{)217}$
$2\overline{)468}$	$6\overline{)726}$
$4\overline{)880}$	$9\overline{)459}$
$3\overline{)627}$	$8\overline{)344}$

Popcorn!

I love popcorn! I like it with a lot of butter and a little bit of salt. I like it so much, I used my allowance to buy a big bag of popcorn kernels. I knew my mom would let me use the salt and butter we have at home.

The bag of popcorn kernels cost me 84¢. I gave the clerk at the store $1.00. How much change did I get back?	One cup of popcorn kernels made four cups of popped corn. There were 8 cups of kernels in the bag. How many cups of popped corn did I get?
I ate two cups of popcorn every night while I watched television. How many cups did I eat in one week?	Three of my friends came over Saturday to play catch. I make a bowl of popcorn for a snack. If the bowl holds 12 cups of popcorn, how much will we each get to eat? (Don't forget to count me when you answer the questions!)
We drank apple juice with our popcorn. We each drank 2 glasses. How much apple juice did we drink?	I started cleaning up the popcorn popper and washing the juice glasses at 4:15. It took me 25 minutes. What time did I finish?

8 ? 5

What Is My Number?

4 ? 2

Problem	Work Area
My number is between 0 and 9. It cannot be divided by 2. It is less than 9 and more than five. What is my number?	
My number is less that 20. It is an odd number. It is not a 2-digit number. It is not the number of sides on a triangle. It is not the number of days in a week., It can be divided by three. What is my number?	
My number is between 20 and 40. It is an odd number. Its digits add up to 8. The largest digit minus the smaller digit is 2. What is my number?	
Make up your own number puzzle.	

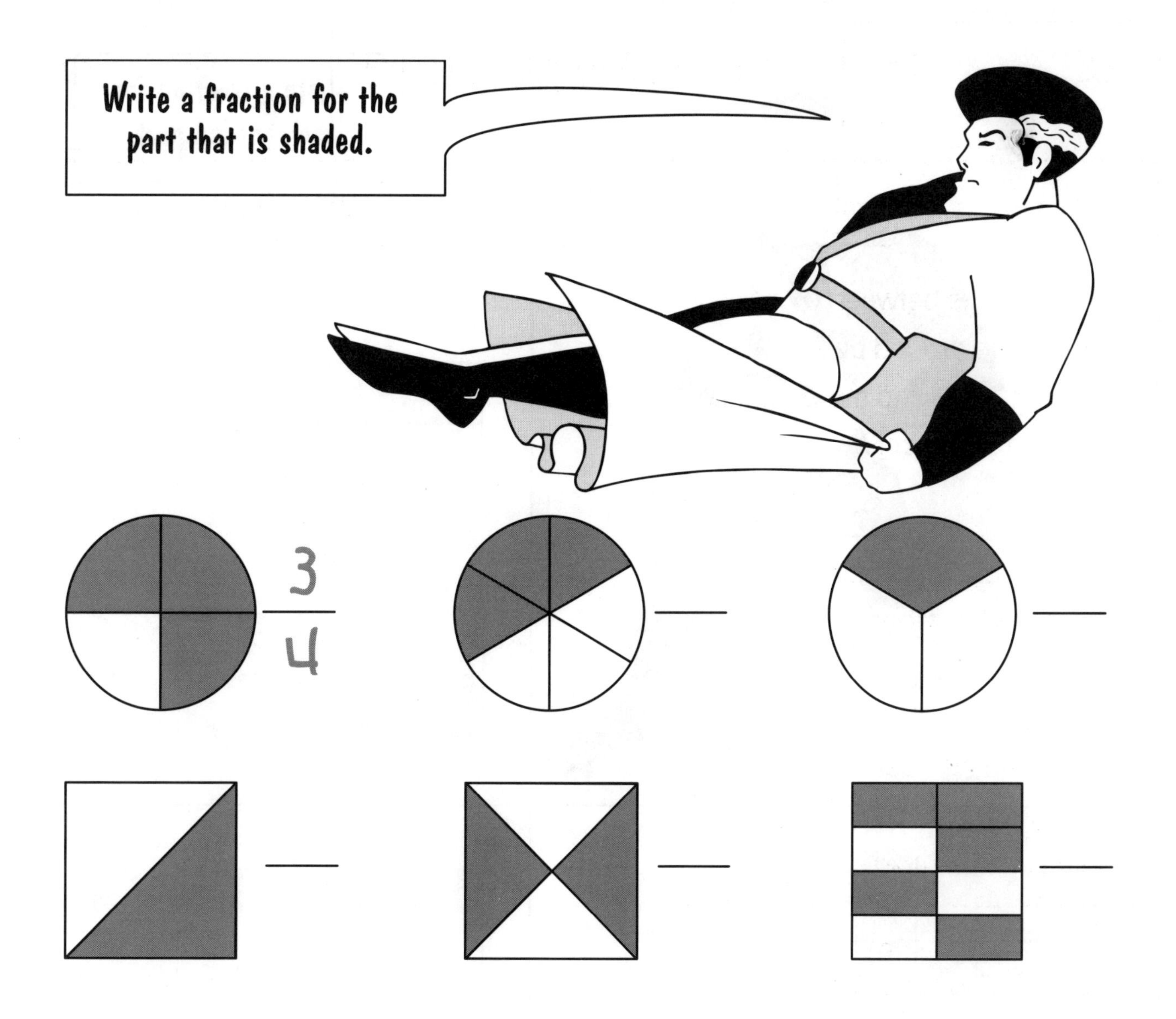

Color parts to show the fraction.

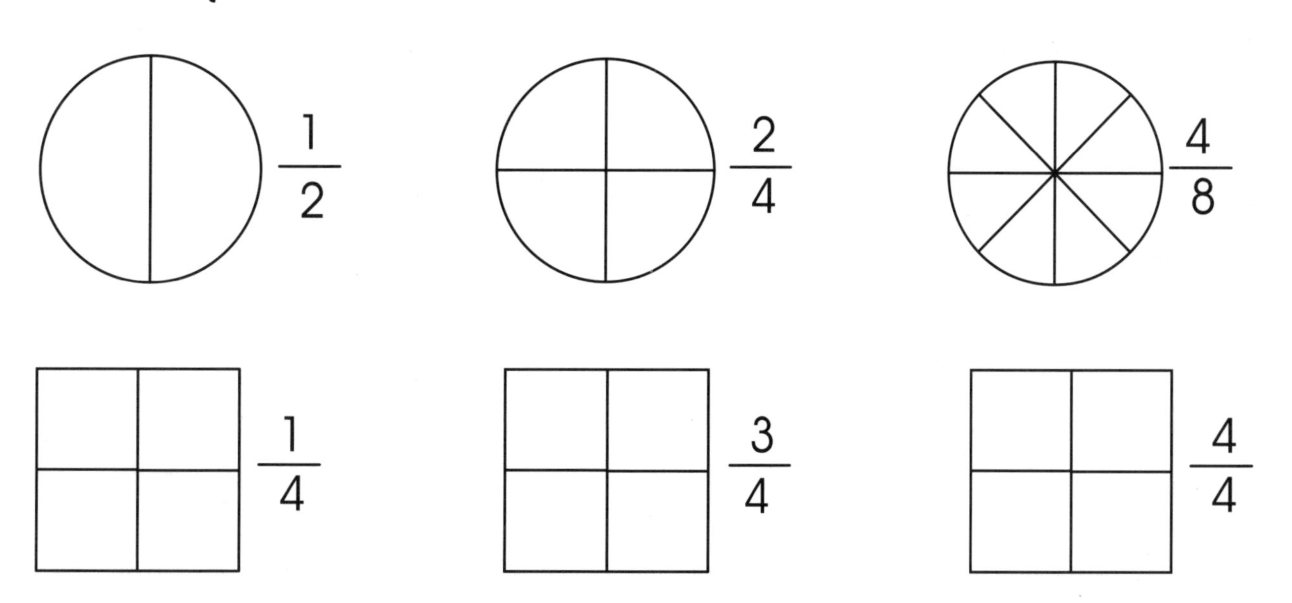

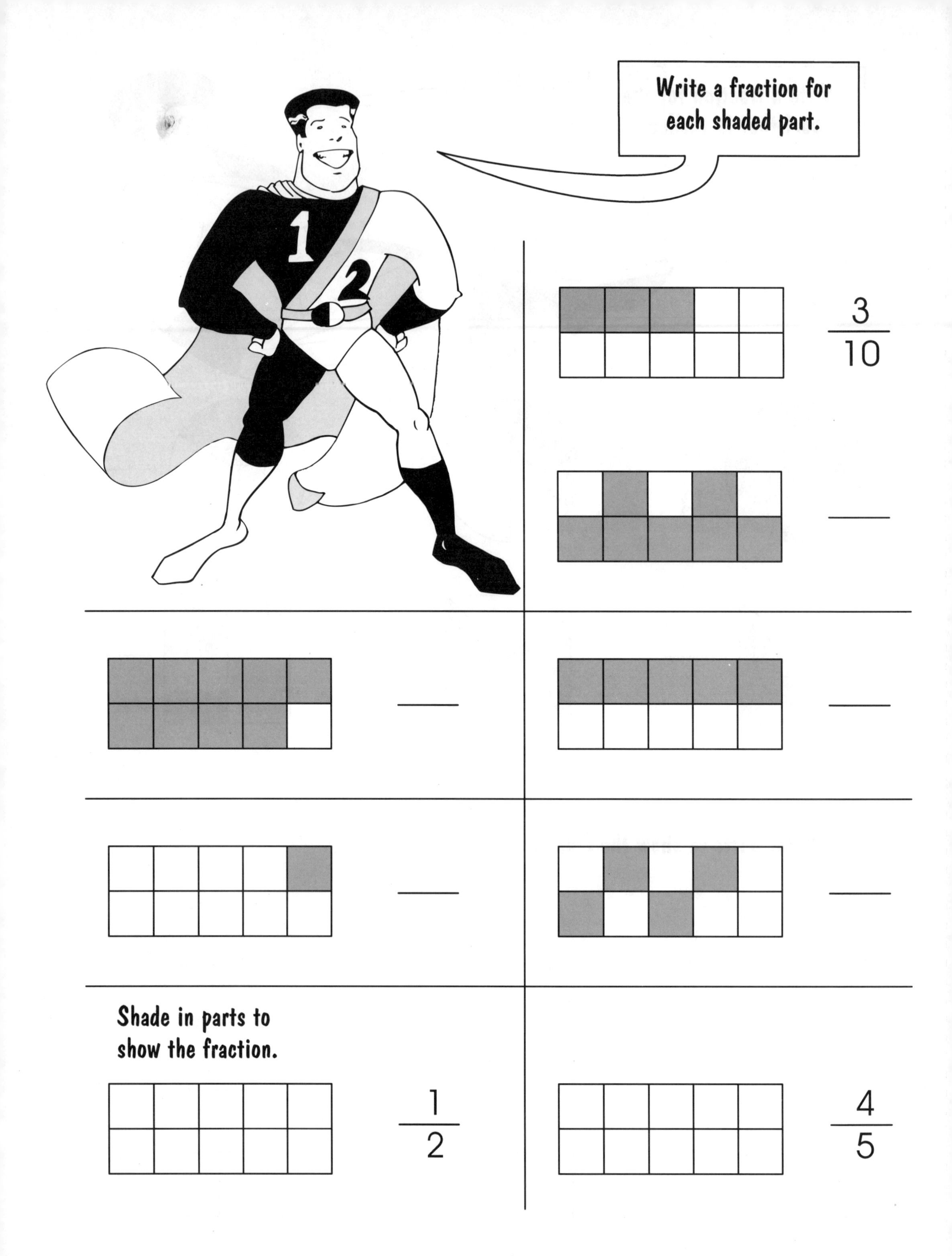
Write a fraction for each shaded part.
1
2
3/10
Shade in parts to show the fraction.
1/2
4/5

Equivalent Fractions

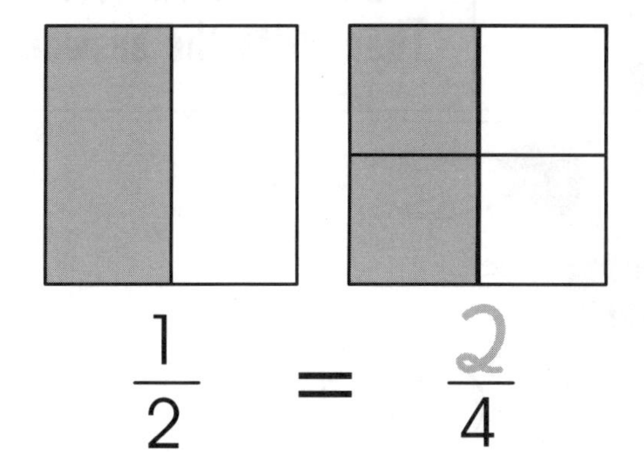

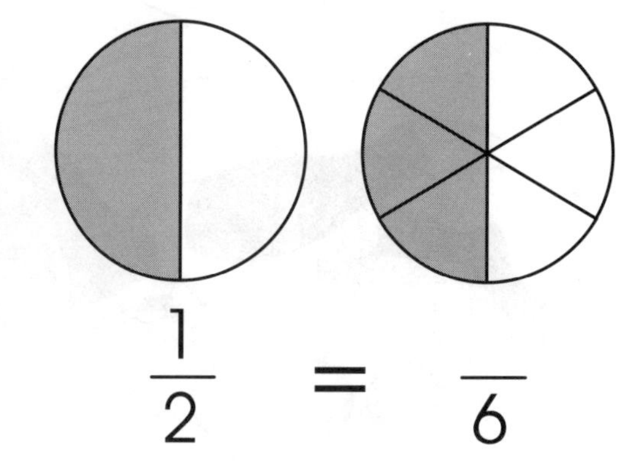

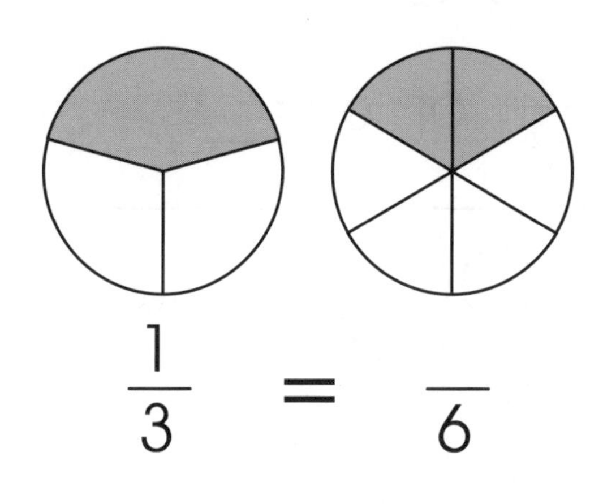

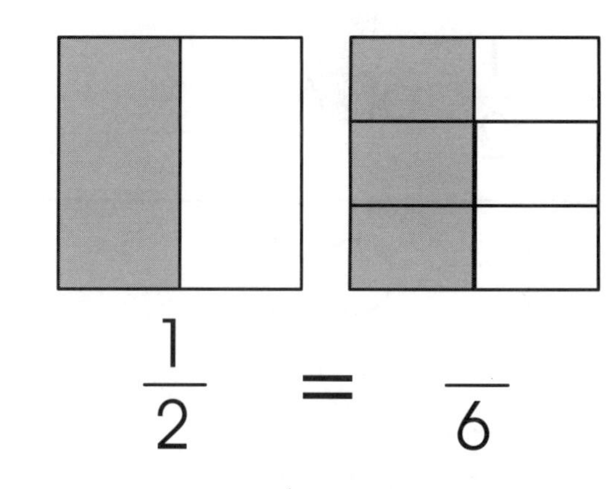

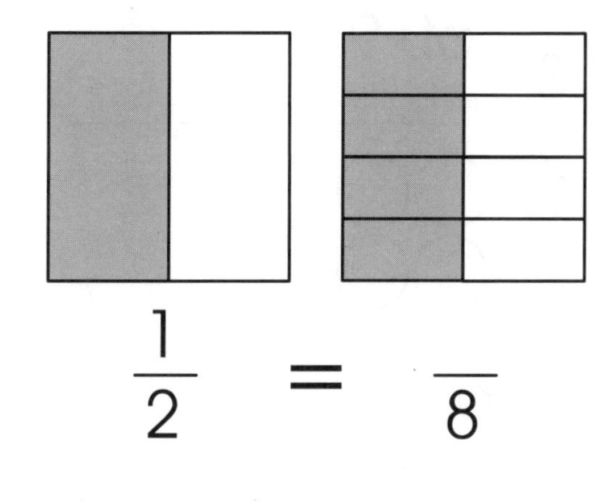

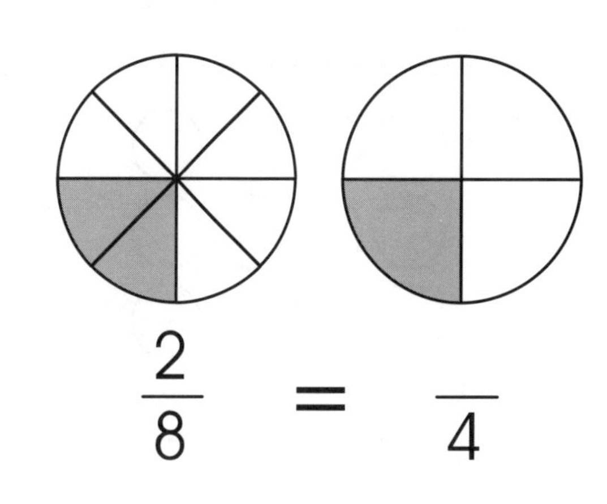

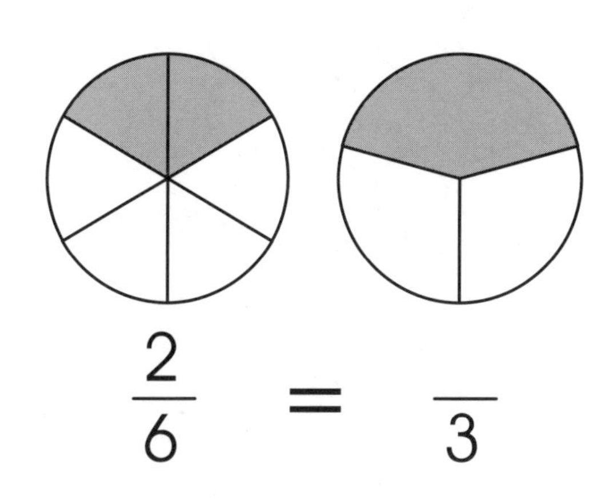

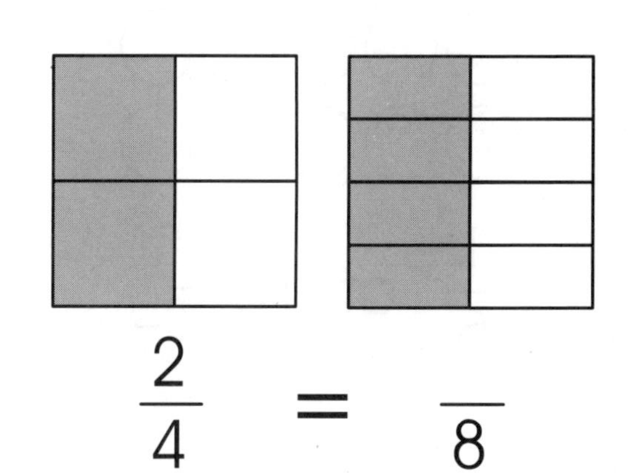

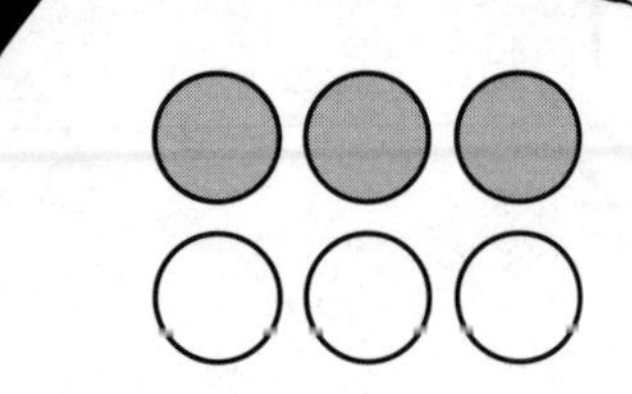

$\frac{1}{2}$ of 6 = 3

$\frac{1}{3}$ of 6 = __

$\frac{1}{2}$ of 4 = __

$\frac{1}{3}$ of 9 = __

$\frac{1}{2}$ of 12 = __

$\frac{1}{8}$ of 8 = __

$\frac{1}{5}$ of 10 = __

$\frac{1}{4}$ of 16 = __

1. Buzzy ate $\frac{1}{3}$ of the pizza. How many slices did he eat?

2. Allie ate $\frac{1}{6}$ of the pizza. How many slices did she eat?

3. I ate $\frac{1}{4}$ of the pizza. How many slices did I eat?

4. What fraction of the pizza is left?

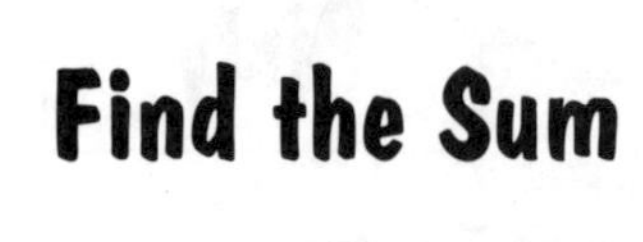

$\frac{2}{5} + \frac{1}{5} = \frac{3}{5}$	$\frac{6}{14} + \frac{5}{14} = \_\_\_$
$\frac{1}{7} + \frac{3}{7} = \_\_\_$	$\frac{5}{10} + \frac{2}{10} = \_\_\_$
$\frac{5}{17} + \frac{2}{17} = \_\_\_$	$\frac{1}{4} + \frac{2}{4} = \_\_\_$
$\frac{4}{11} + \frac{5}{11} = \_\_\_$	$\frac{4}{6} + \frac{1}{6} = \_\_\_$
$\frac{1}{3} + \frac{1}{3} = \_\_\_$	$\frac{2}{8} + \frac{1}{8} = \_\_\_$
$\frac{3}{9} + \frac{4}{9} = \_\_\_$	$\frac{2}{5} + \frac{2}{5} = \_\_\_$

$\frac{3}{4} - \frac{2}{4} = \frac{1}{4}$	$\frac{6}{7} - \frac{4}{7} = \square$
$\frac{10}{18} - \frac{7}{18} = \square$	$\frac{9}{11} - \frac{1}{11} = \square$
$\frac{18}{26} - \frac{11}{26} = \square$	$\frac{4}{5} - \frac{3}{5} = \square$
$\frac{2}{3} - \frac{1}{3} = \square$	$\frac{13}{14} - \frac{2}{14} = \square$
$\frac{9}{21} - \frac{5}{21} = \square$	$\frac{7}{8} - \frac{4}{8} = \square$
$\frac{7}{9} - \frac{2}{9} = \square$	$\frac{9}{12} - \frac{2}{12} = \square$

Fractions greater than 1 can be written in two ways: as a fraction; as a mixed number.

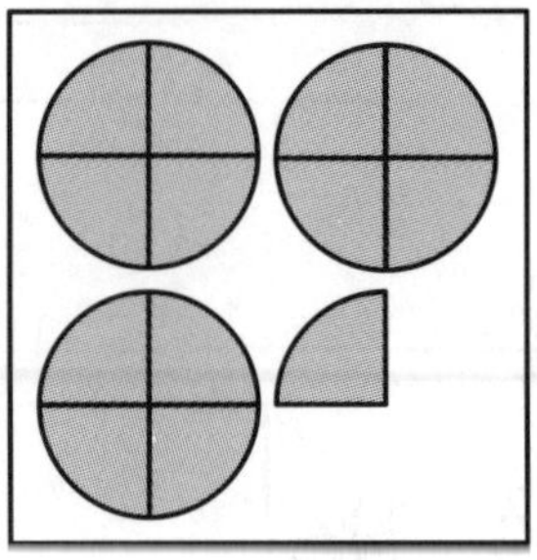

$\frac{13}{4}$ $3\frac{1}{4}$

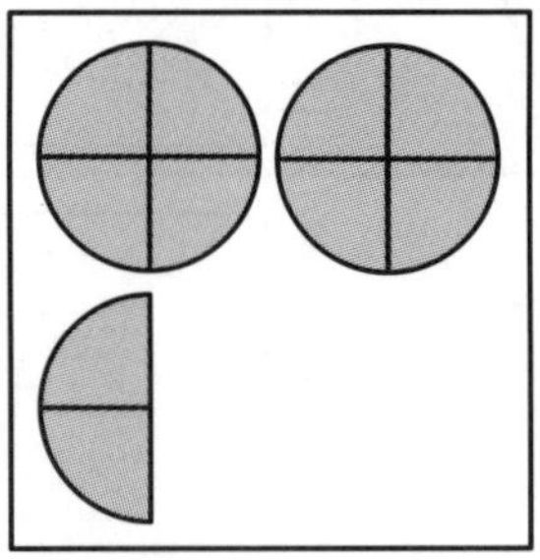

____ ____

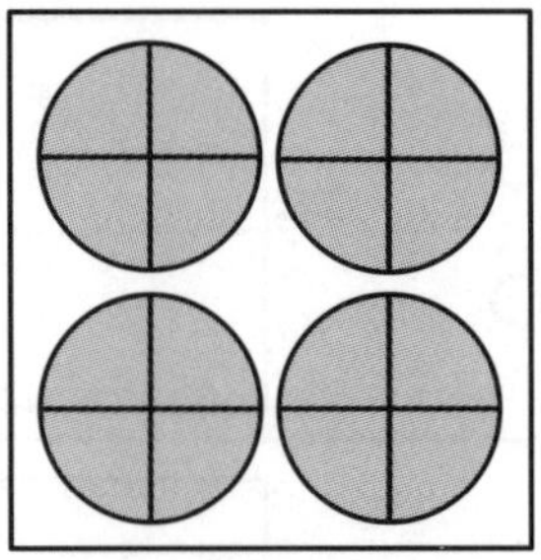

____ ____

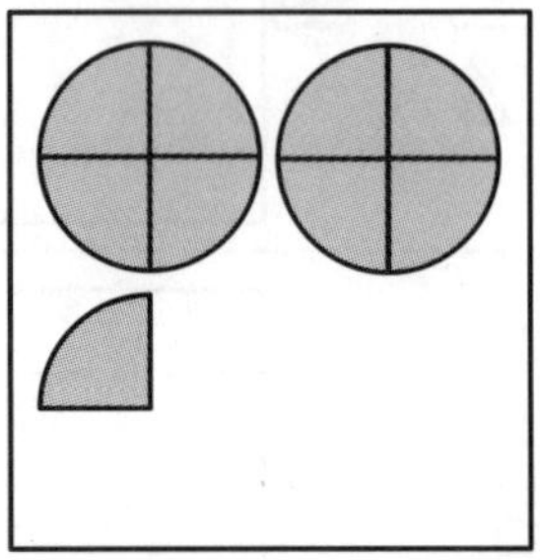

____ ____

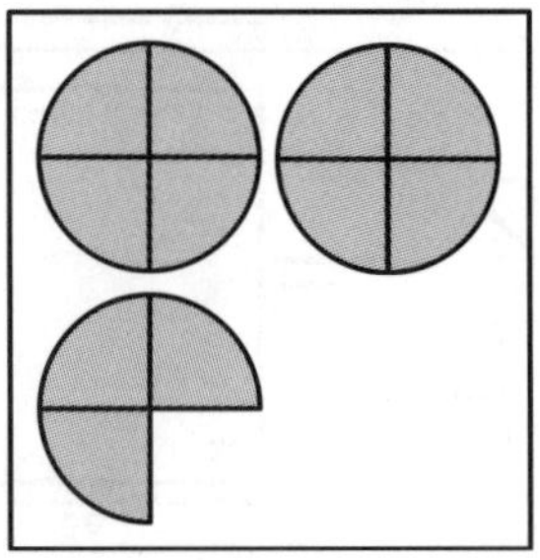

____ ____

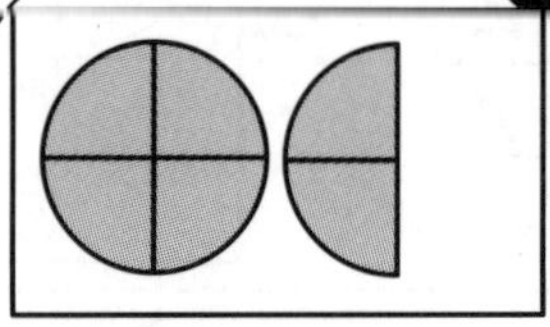

____ ____

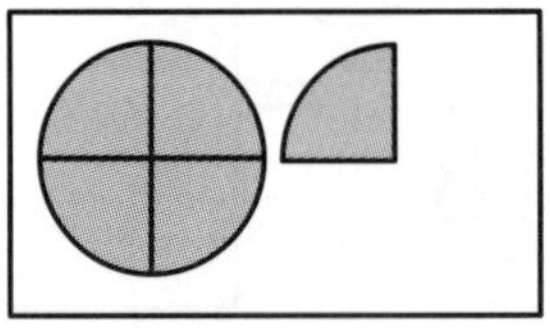

____ ____

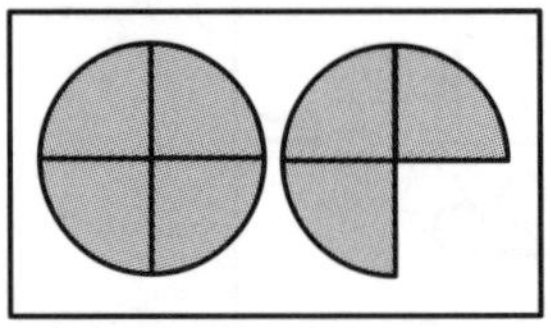

____ ____

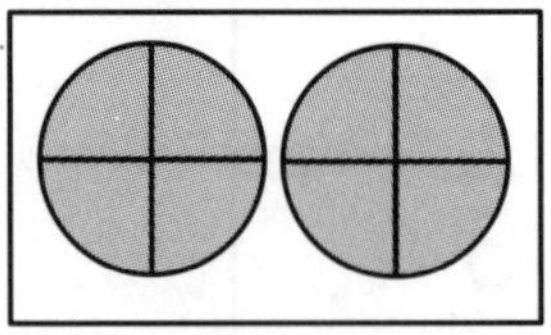

____ ____

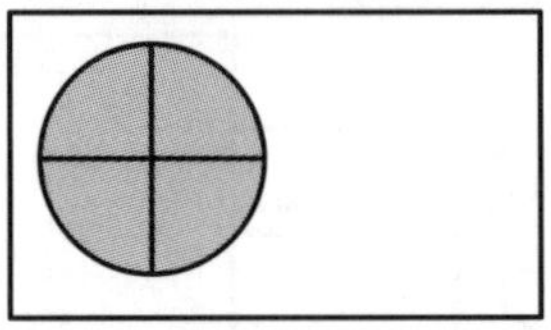

____ ____

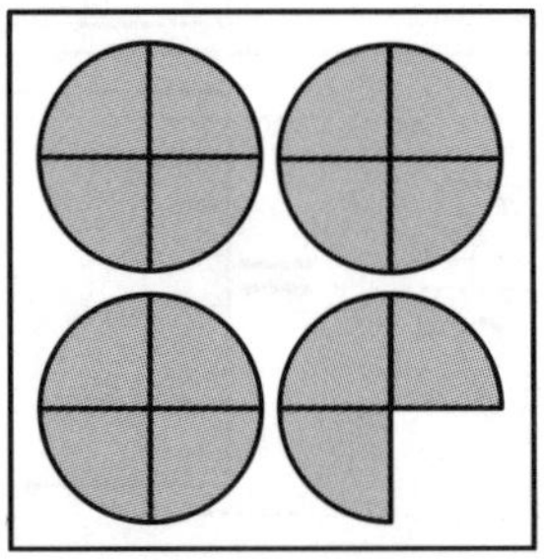

____ ____

____ ____

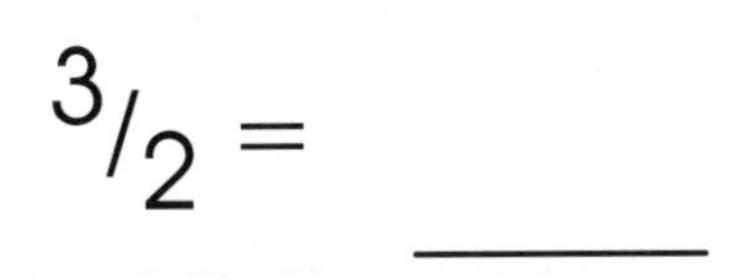

$^3/_2 =$ ______ $^3/_3 =$ ______

$^5/_4 =$ ______ $^6/_4 =$ ______

$^8/_4 =$ ______ $^{11}/_5 =$ ______

$^7/_2 =$ ______ $^9/_6 =$ ______

$^5/_3 =$ ______ $^6/_2 =$ ______

$^7/_5 =$ ______ $^4/_4 =$ ______

Write as an improper fraction.

An improper fraction has a bigger **numerator** than a **denominator.**

For example: $\frac{6}{5}$

Think: $1 = \frac{4}{4}$; $\frac{4}{4} + \frac{1}{4} = \frac{5}{4}$ — $1\frac{1}{4} = \frac{5}{4}$	Think: $5 = \frac{30}{6}$ — $5\frac{3}{6} =$
$2\frac{1}{2} =$	$3 = \frac{12}{}$
$3\frac{2}{3} =$	$1\frac{1}{2} =$
$2 = \frac{}{4}$	$4\frac{3}{7} =$
$4\frac{3}{5} =$	$3\frac{2}{5} =$
$1 = \frac{}{6}$	$2\frac{1}{4} =$
$3\frac{3}{4} =$	$4 = \frac{16}{}$

Write the fraction and decimal for the shaded part of these shapes.

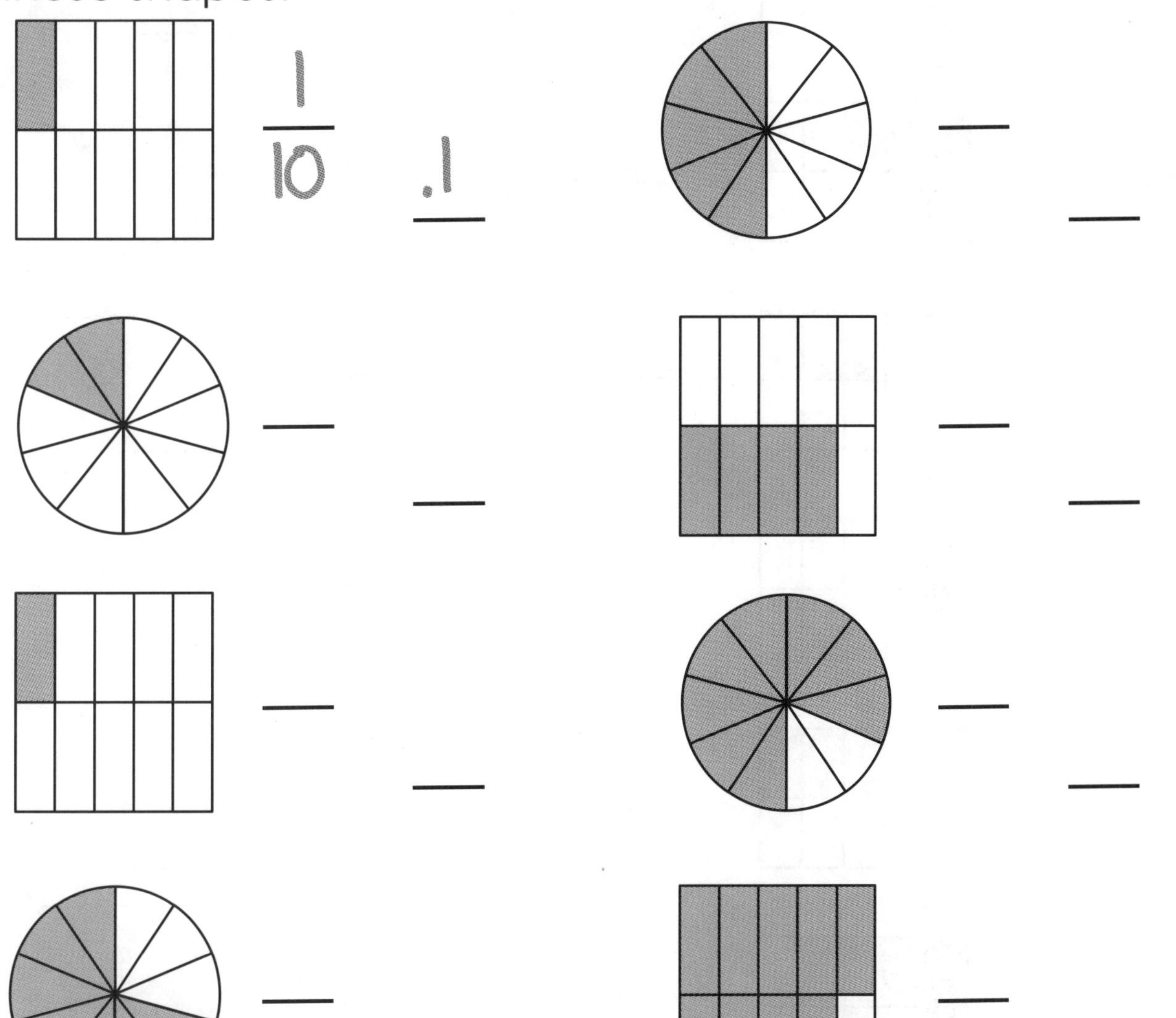

Fractions and decimals are both ways we name parts of a whole.

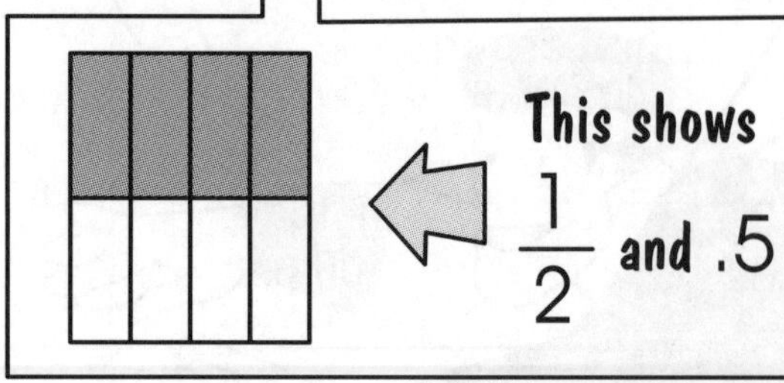

Write the fraction and the decimal name for each of these.

= $\frac{4}{10}$ = .4

= $\frac{3}{10}$ = .3

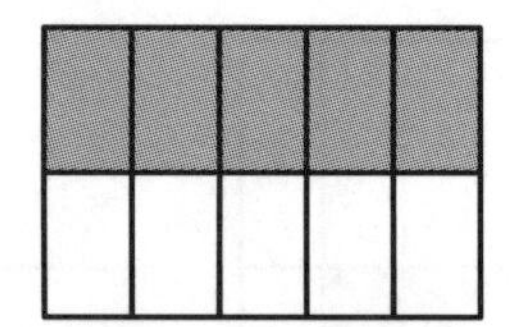 = ________ = ________

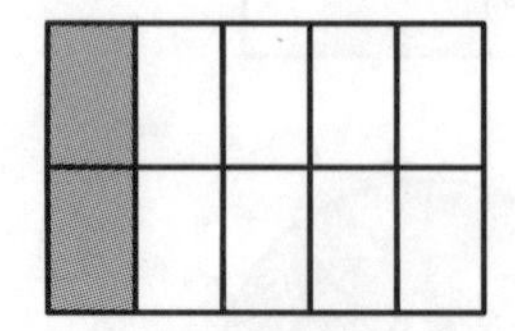 = ________ = ________

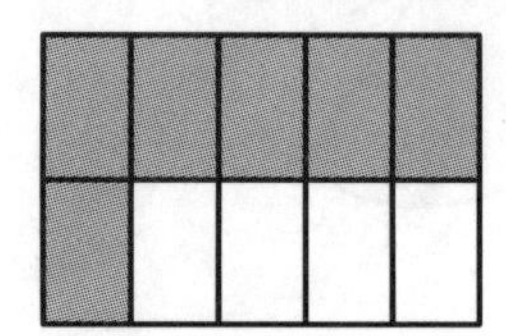 = ________ = ________

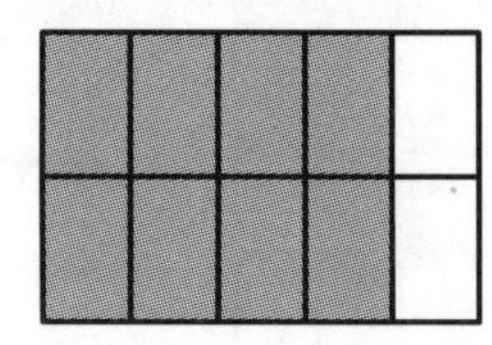 = ________ = ________

That Should Take About...

Directions: Use your estimation skills to solve a problem. Circle the correct answer along with the letter beside it. Each time the number of the problem appears at the bottom of the page, write the letter of the answer above it. When you are all done, decode the answer to the riddle.

Choose the best estimate for the following:

1. The time it takes to wash and dry your hands.
V. 1 second **R.** 1 minute **L.** 1 hour

2. The time it takes to view a movie at the movie theater.
A. 2 hours **C.** 20 minutes **T.** 2 days

3. The time it takes to say your name.
K. 2 minutes **E.** 2 seconds **D.** 20 seconds

4. The time it takes to brush your teeth.
M. 3 seconds **S.** 30 minutes **Y.** 3 minutes

5. The time it takes to dial a telephone.
C. 5 seconds **W.** 5 minutes **P.** 5 hours

6. The time it takes to blow out a birthday candle.
A. 1 minute **T.** 1 second **E.** 1 hour

7. The amount of time the average person sleeps at night.
B. 8 minutes **Y.** 8 seconds **O.** 8 hours

8. The amount of time it takes to read and solve this problem.
S. 30 hours **B.** 30 minutes **V.** 30 seconds

9. The amount of time it takes to sneeze.
A. 2 hours **K.** 2 seconds **O.** 2 minutes

10. The amount of time it takes to read this sentence.
D. 4 seconds **N.** 40 seconds **S.** 4 minutes

A famous frog hero.

___	___	___	___		___	R	___	___	___	___	___	___
10	2	8	4		5	1	7	2	9	3	6	6

Elapsed Time

Elapsed time is the amount of time between one event and another. For example, if a parade began at 11:00 A.M. and ended at 1:20 P.M., the elapsed time between the start and end of the parade is 2 hours 20 minutes.

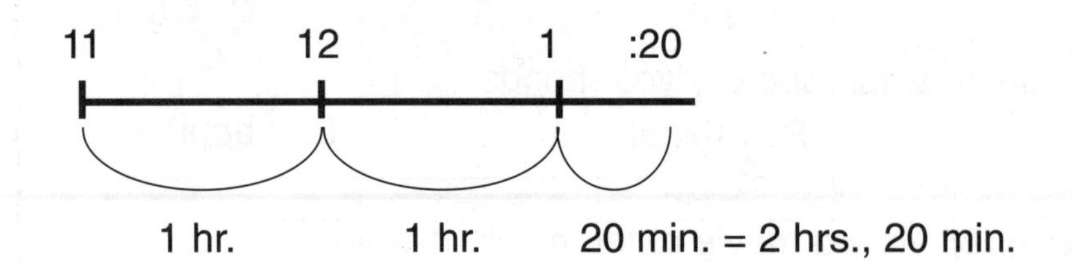

Directions: Find the answer in the answer column. Each time the number of the problem appears at the bottom of the page, write the letter of the answer above it. When you are all done, decode the answer to the silly saying.

1. 240 minutes = ______ hours.

2. 72 hours = ______ days.

3. 1.5 hours = ______ minutes.

4. 300 seconds = ——- minutes.

5. What time is five hours after 9:00 A.M.?

6. What time is 2 hours and 10 minutes before 3:00 P.M.?

7. What time is 1 hour and 45 minutes before 12:30 P.M.?

8. How much time is there between 5:00 A.M. and 10:15 A.M.?

9. How much time is there between 11:30 A.M. and 3:00 P.M.?

10. How much time is there between 11:45 P.M. and 4:10 A.M.?

11. How much time is there between 3:15 A.M. and 3:30 P.M.?

12. The plane left Miami at 8:30 A.M. and flew non-stop to Baltimore. The trip took 2 hours and 55 minutes. At what time did the plane arrive in Baltimore?

Answers:

S 10:45 A.M.
N 90
B 3 hrs 30 min
L 3
A 12:50 P.M.
O 5 hrs 15 min
E 4 hrs 25 min
R 4
V 12 hrs 15 min
F 11:25 A.M.
H 2:00 P.M.
T 5

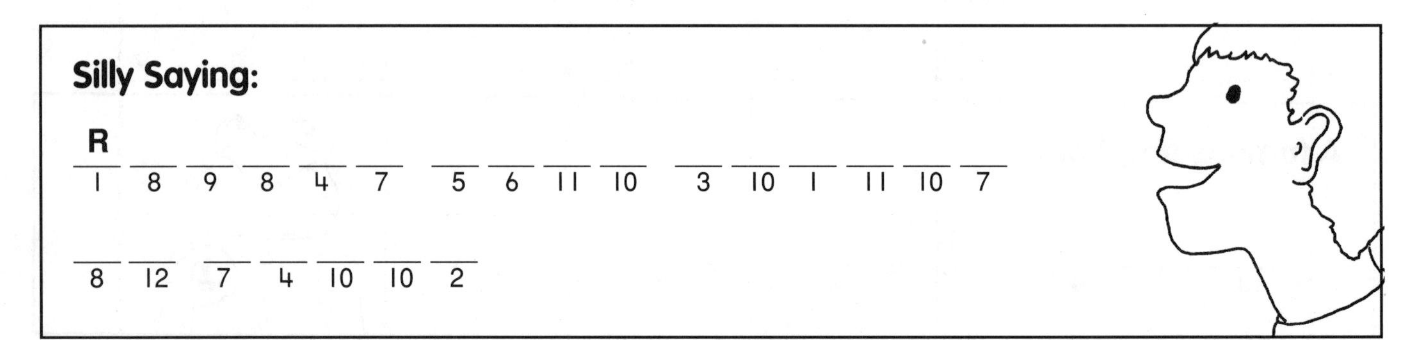

How Many Ways Can You Make $1.00?

Use this money table to help you figure out nine of the combinations of coins that can be put together to make exactly $1.00.

penny	nickel	dime	quarter	half dollar	$
100					=$1.00
	5		3		=$1.00

Making Change

Directions: Calculate how much change will be returned for each purchase. Find the answer in the answer column. Each time the number of the problem appears at the bottom of the page, write the letter of the answer above it. When you are all done, decode the answer to the dieter's motto.

Key: P = penny, N = nickel, D = dime, Q = quarter, $1 = one dollar

1. Cost $4.60. Paid with $5 bill. Change = **B**

2. Cost $18.20. Paid with $20 bill. Change = ?

3. Cost $7.95. Paid with $10 bill. Change = ?

4. Cost $.75. Paid with $1 bill. Change = ?

5. Cost $.35. Paid with two quarters. Change = ?

6. Cost $.70. Paid with $1 bill. Change = ?

7. Cost $19.95. Paid with $20 bill. Change = ?

8. Cost $2.90. Paid with three $1 bills. Change = ?

9. Cost $8.19. Paid with $10 bill. Change = ?

10. Cost $.84. Paid with $1 bill. Change = ?

11. Cost $1.50. Paid with two $1 bills. Change = ?

12. Cost $.05. Paid with a quarter. Change = ?

13. Cost $.88. Paid with $1 bill. Change = ?

14. Cost $18.95. Paid with $20 bill. Change = ?

15. Cost $7.95. Paid with $10 bill. Change = ?

16. Cost $.65. Paid with $1 bill. Change = ?

Answers

E	Q
M	N
D	N, $1, $1
L	Q, Q
U	P, P, D
S	N, D
I	N, Q, Q, Q, $1
A	D
O	N, Q
T	D, D
B	N, Q, D
C	P, N, D
H	N, $1
D	P
J	P, N, Q, Q, Q, $1
N	D, Q

Dieter's Motto

___ ___ ___ ___   ___ ___ ___ ___ ___   ___ ___ ___
3 6 16 12   10 6 13 16 12   12 14 4

___ ___ ___ ___ ___ ___ ___   ___ ___   ___ ___ ___   ___ ___ B ___ ___
7 2 16 13 12 4 5   8 12   12 14 4   12 8 1 11 4

___ ___ ___ ___   ___ ___ ___   ___ ___ ___ ___ ___ ___ ___
9 13 5 12   12 14 4   5 4 10 6 16 15 5

Adding and Subtracting Money

When adding and subtracting money, be sure to align the decimal points before performing the operation with the numbers. For example, to add $1.46 and $12.04, align the decimal points, then add the numbers. The final step is to bring the decimal point straight down into the answer.

Directions: Use addition or subtraction to solve a problem. Find the answer in the answer column. Each time the number of the problem appears at the bottom of the page, write the letter of the answer above it. When you are all done, decode the silly saying.

Hint: Rewrite the problems like this:

```
$ 1.41
$  .34
------
$ 1.75
```

1. $1.41 + $.34 =
2. $8.56 + $7.04 =
3. $4.75 - $2.40 =
4. $2.85 - $1.69 =
5. $18.21 - $17.95 =
6. $4.36 + $2.79 =
7. $9.48 + $8.53 =
8. $8.51 - $3.84 =
9. $72.85 + $.95 =
10. $4.75 + $.79 =
11. $0.37 + $1.94 =
12. $74.92 - $2.12 =
13. $20.62 + $1.59 =
14. $72.90 - $71.60 =

Answers:

- **S** $2.31
- **H** $1.75
- **U** $18.01
- **E** $2.35
- **D** $73.80
- **C** $.26
- **I** $22.21
- **Y** $15.60
- **N** $1.30
- **O** $4.67
- **A** $7.15
- **M** $5.54
- **L** $1.16
- **R** $72.80

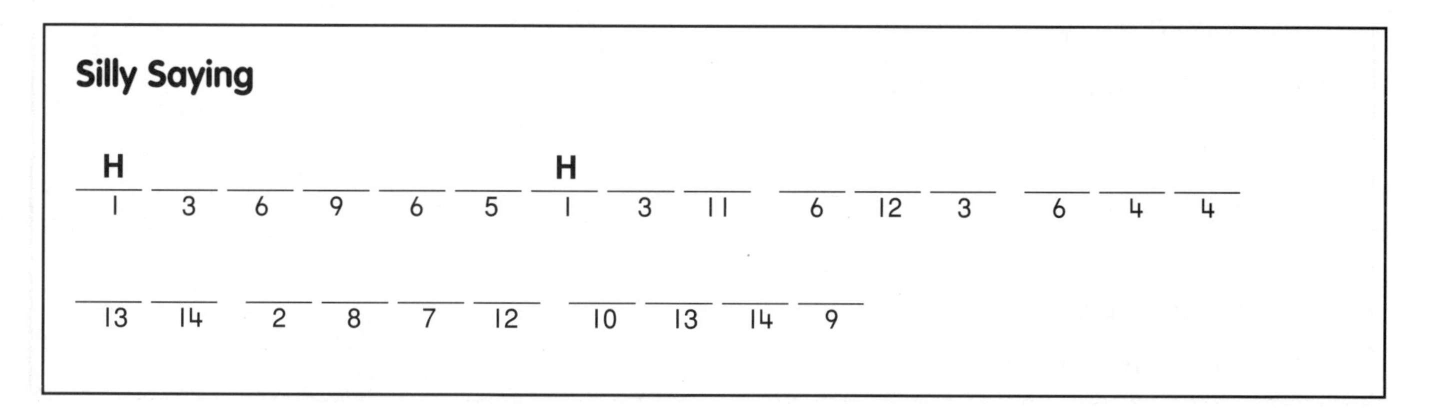

Building a Tree House

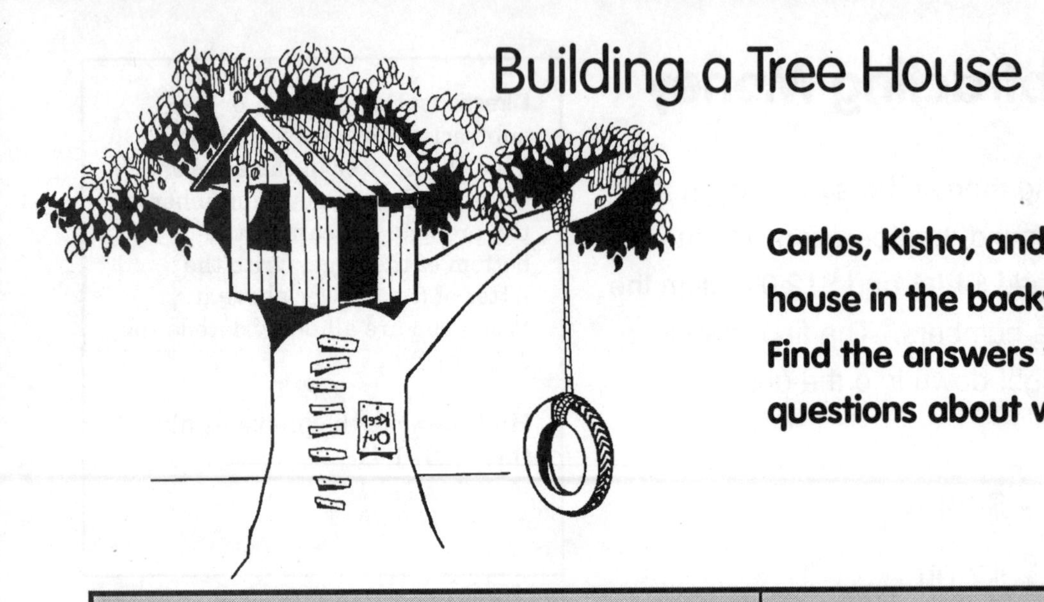

Carlos, Kisha, and Otto built a tree house in the backyard.
Find the answers to these questions about what they did.

Problems	Work Space
Carlos, Kisha, and Otto needed money to buy materials to build the tree house. Carlos did jobs to earn \$18.75. Kisha collected cans and bottles to get \$9.58. Otto saved his allowance until he had \$10.50. How much money did they have?	
They went to the lumberyard and spent \$26.34 for wood, nails, and paint. How much money did they have left?	
They worked every Saturday for 4 weeks. Each Saturday they worked from 9:30 'till 4:00. They stopped for an hour for lunch each day. How much time did they spend building the tree house?	

Using Terms and Tools

It is important to know the vocabulary and the tools of measurement. You can find out how much you already know by filling in the statement below with the best measurement term or tool at the bottom of the page.

Directions: Each time the answer appears, cross it out along with the letter above it. When you are all done, the remaining letters will spell out the answer to the riddle.

1. The temperature of air or water is measured in ________.

2. Hours and minutes are a measure of ________.

3. ________ is sometimes measured in grams and kilograms.

4. Cubic inches are used to describe the ________ of a container.

5. The ________ of a playground is measured in square feet.

6. The ________ of a container can be measured in liters or gallons.

7. Yards or meters can be used to measure the ________ between two objects.

8. A ________ can tell you how much an object weighs.

9. You can use a ________ to find the number of days until your birthday.

10. A ________ can be used to find the length of a pencil.

11. The temperature of water can be measured with a ________.

12. A ________ can tell you how long someone can hold his or her breath.

What occurs twice in every day, four times in every week, but only once in a year?

C	R	S	E	L	T	W	M	A	L	I	D	E
Mass	Distance	Degrees	Kilometer	Calendar	Capacity	Thermometer	Area	Scale	Volume	Time	Ruler	Clock

Comparing Customary Units of Length

Data:

12 inches = 1 foot

3 feet = one yard

1,760 yards = 1 mile

5, 280 feet = 1 mile

Directions: Draw a straight line between a measurement on the left and its equivalent measurement on the right. Correct responses will cross number and a letter. Each time the number appears at the bottom of the page, write the letter above it. When you are all done, decode the answer to the riddle.

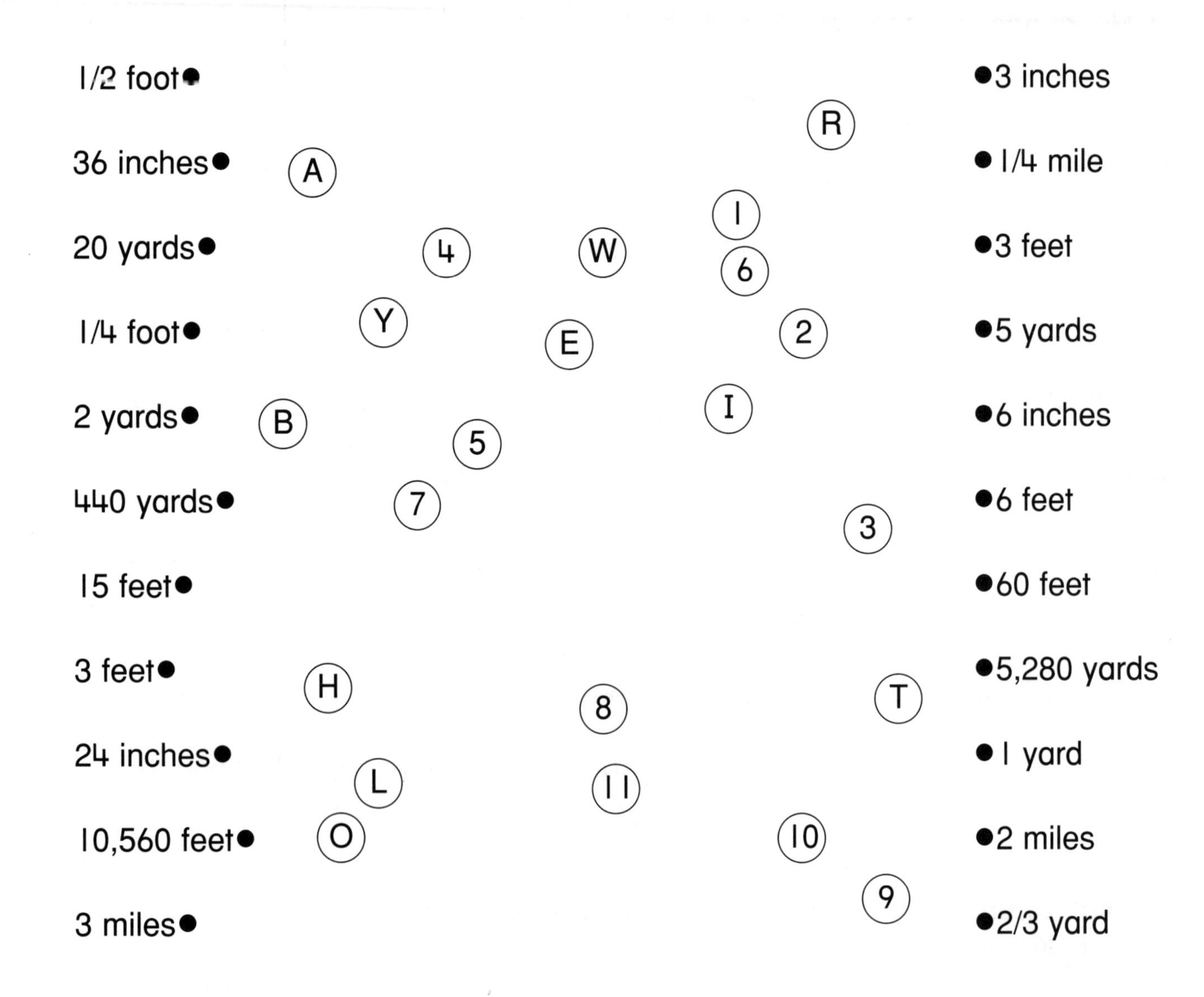

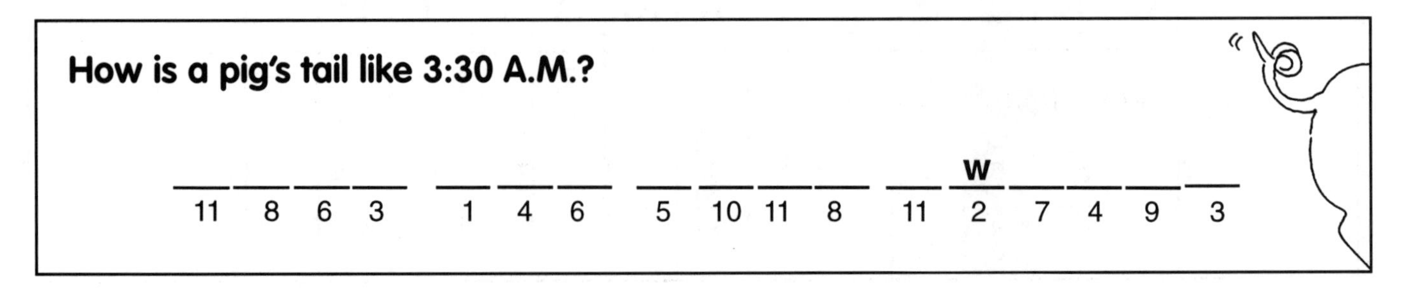

Determining Perimeters

Directions: Use an inch ruler to measure the perimeter of a shape to the nearest quarter inch. Each time the answer appears at the bottom of the page, cross it out along with the letter above it. When you are all done, unscramble the remaining letters to spell out the answer to the riddle.

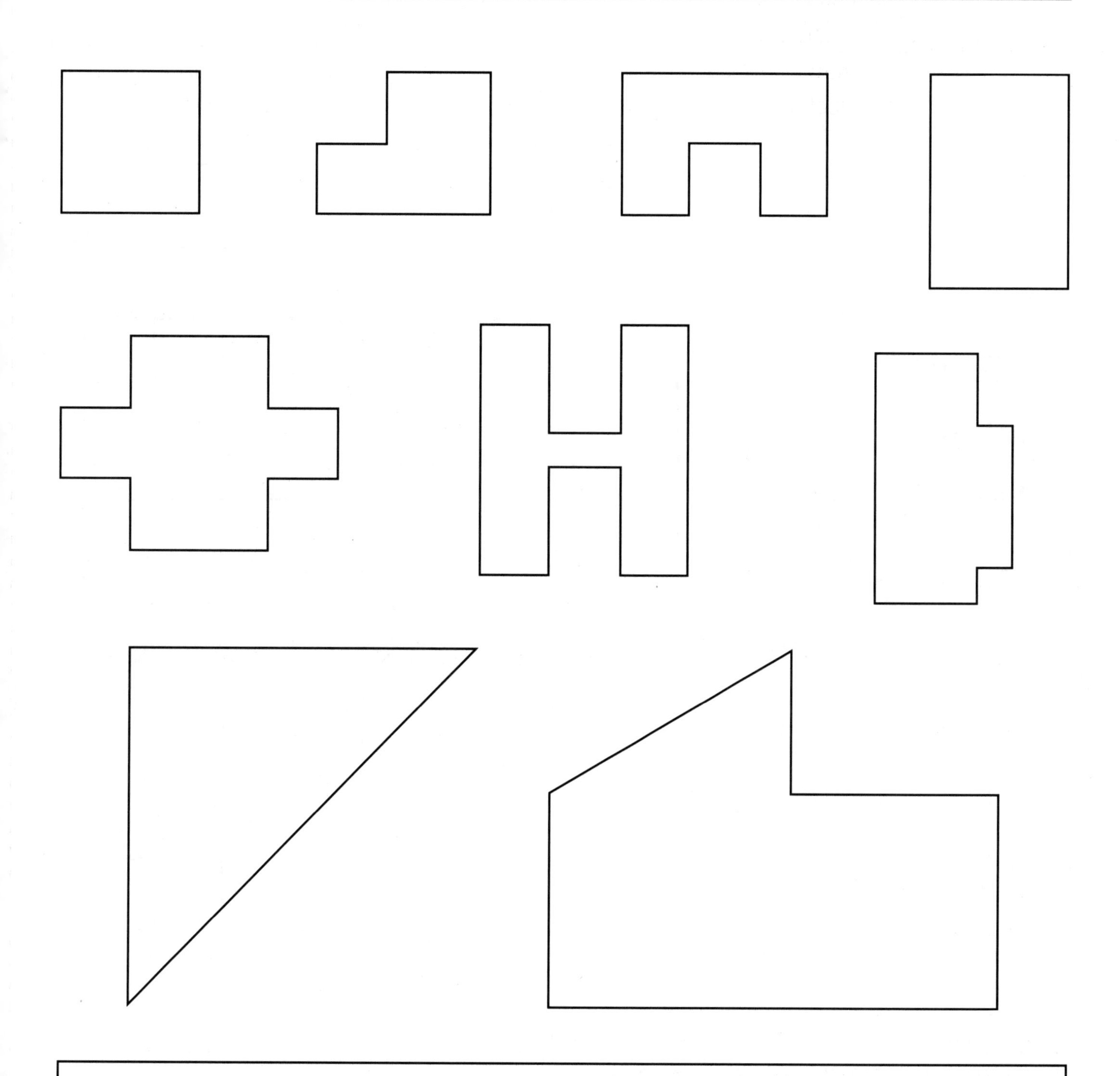

A banker's three favorite letters

M	C	A	O	N	S	E	H	I	L	Y	U
5"	4 1/2"	7"	3"	4"	5 1/2"	9 1/2"	6"	9"	10 3/4"	8 1/2"	2 1/2"

Determining Area

Definition: Area is the number of square units that are needed to completely cover a surface.

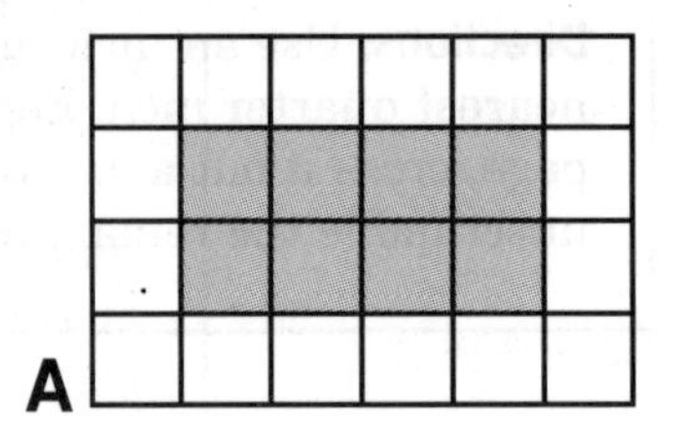

The area of the shaded portion of shape A is 8 square units.

The area of the shaded portion of shape B is also 8 square units. Notice that the half units can be added together to make a whole unit.

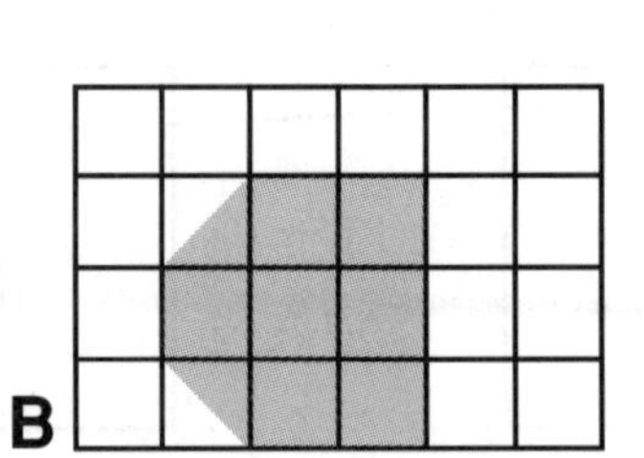

Directions: Calculate the area of each letter in square units. Each time the area appears at the bottom of the page, write the letter above it. When you are all done, decode the answer to the riddle.

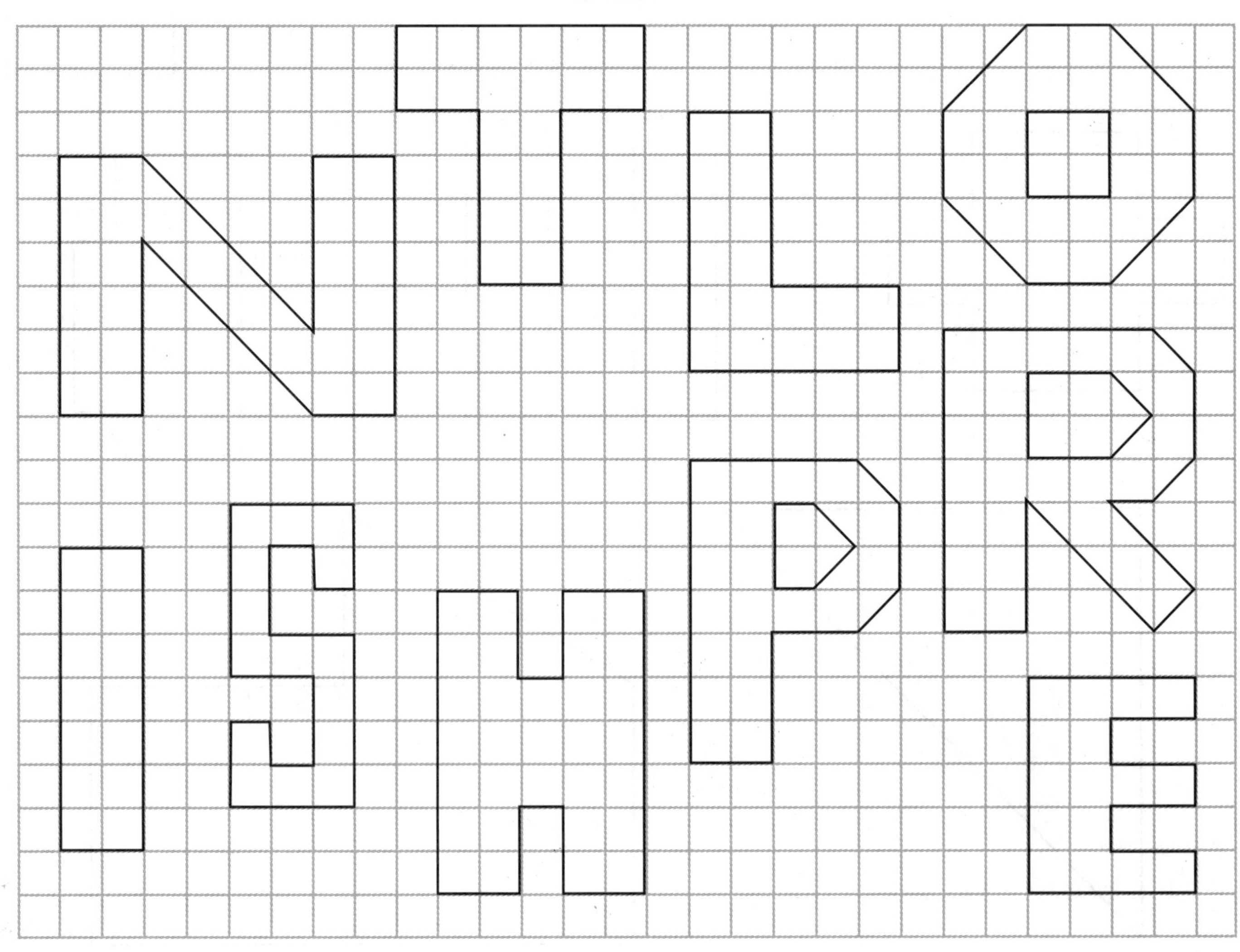

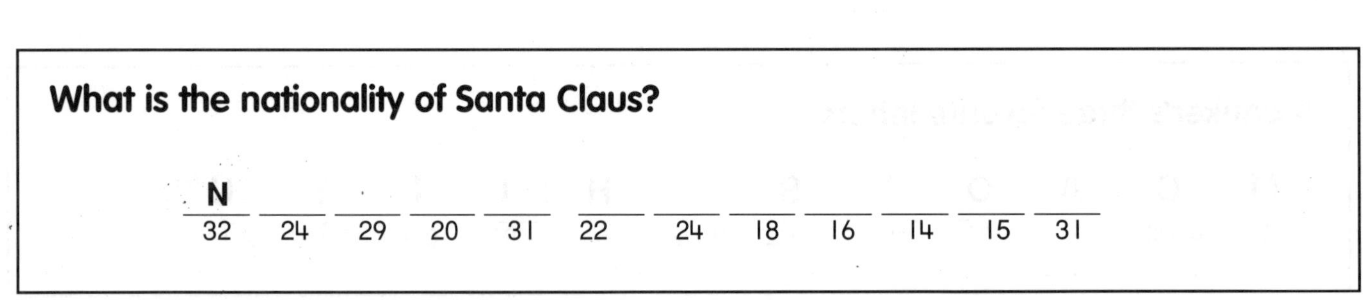

What is the nationality of Santa Claus?

N											
32	24	29	20	31	22	24	18	16	14	15	31

Polygons and Polyhedrons

Definition: A polyhedron is a closed, solid (three dimensional) figure with four or more flat sides.

Directions: The descriptions of several polygons and polyhedrons are listed below. Name the polygon or polyhedron that is being described.

Find the answer box at the bottom of the page. Each time the answer appears, write the letter of the problem above it. When you are done, decode the message on the sign.

O. I am a solid figure. I have six flat sides. They all look exactly the same.

What Am I?

C. I am a flat figure. A flag, a box top, and a piece of paper all have my shape.

What Am I?

E. I am a flat figure. I have three sides.

What Am I?

A. I am a flat shape. If you traced me 6 times, you could use me to make a cube.

What Am I?

W. I am a flat figure. I am found on one side of a cone. I am also found on one side of a cylinder.

What Am I?

S. I am a solid figure. If you traced my sides, you would draw four triangles and a square.

What Am I?

M. I am a solid figure. One end of me is flat and the other end comes up to a point.

What Am I?

E. I am a flat figure. I have six straight lines.

What Am I?

Sign at a lumberyard . . .

C	O	M	E	S	E	E	C	O	M	E	S	A	W
Rectangle	Cube	Cone	Triangle	Pyramid	Triangle	Triangle	Rectangle	Cube	Cone	Triangle	Pyramid	Square	Circle

Symmetry in Art

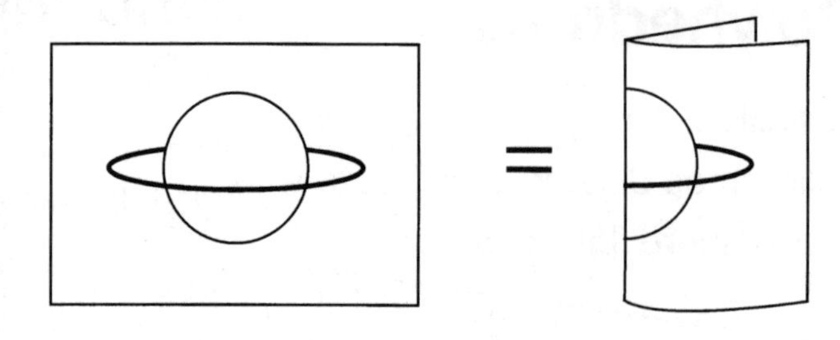

If an object or drawing is folded along a line of symmetry, the two halves will match exactly. For example, the line of symmetry of any circle is a diameter of the circle. When the circle is folded in half on a diameter, both halves of the circle are the same size and shape.

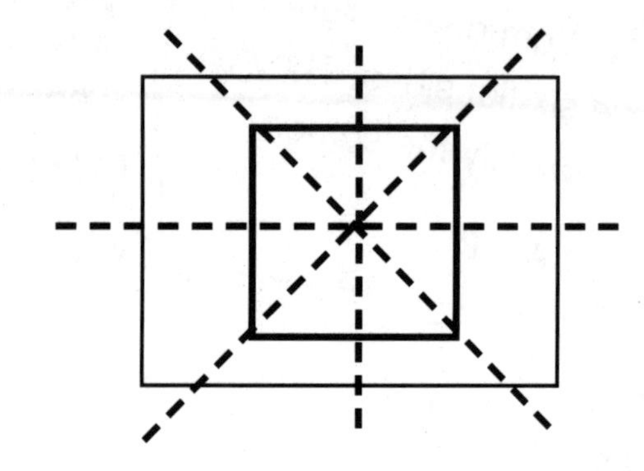

A square has 4 possible lines of symmetry as shown here. If the square is folded on any of the lines of symmetry, the two halves will match up exactly.

Sometimes you can find lines of symmetry in pictures of objects or people. If you can, you might be able to use the picture as part of a fun art project.

Here's how.

1. Look through old magazines for pictures of objects or people. Find one or two that have an obvious line of symmetry. (A picture of someone's face looking straight into the camera is excellent for this activity.) Get permission first, then cut out the picture.

2. Using a pencil and ruler, draw in the line of symmetry.

3. Cut the picture in half along the line of symmetry and paste one part onto a piece of drawing paper as shown in this illustration. Put the other half of the picture aside for now.

4. Remember that a line of symmetry divides an object or picture into two congruent parts and one part is a mirror image of the other. Using your observation skills and the portion of the picture pasted to the page, try to draw in the portion of the picture that is missing. When you are done, compare your drawing to the half of the picture you set aside. How did you do?

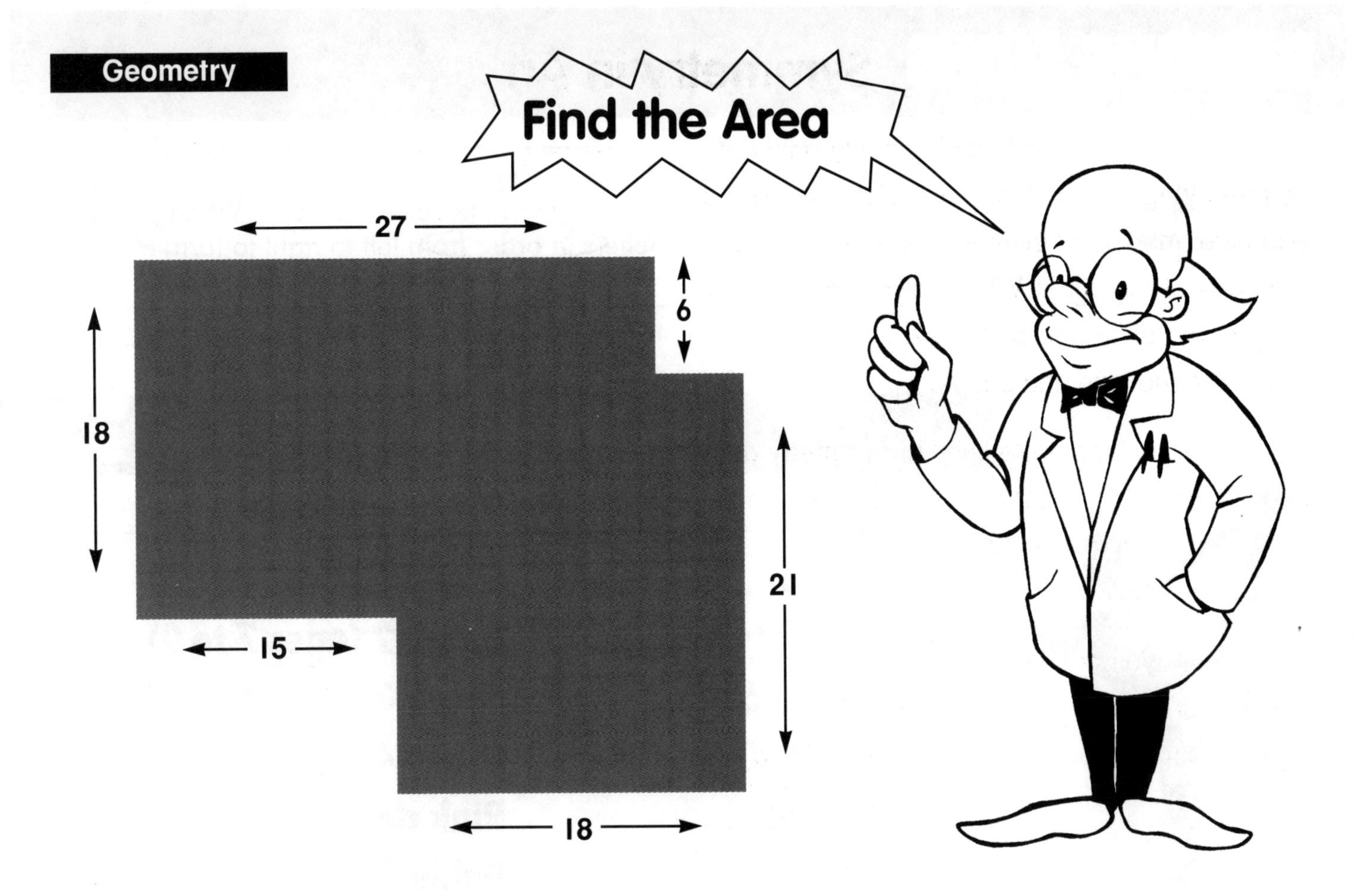

Study this picture. Use logical thinking to figure out the missing lengths.
Then figure out the area of its shape.

The area is __________ square units.

Work space:

READING SIGNS

Use the information on the sign to solve a problem. Each time the answer appears at the bottom of the page, cross it out along with the letter above it. When you ar all done, write down the remaining letters in order from left to right to form a word that answers the riddle.

Ice Skating Lessons

Time	Group
9:00 - 9:45	Group 1 (ages 4 to 6)
9:45 - 10:30	Group 2 (ages 7 to 9)
10:30 - 11:15	Group 3 (ages 10 to 11)
11:15 - 12:00	Group 4 (ages 12 to adult)
12:00 - 1:00	Rink closed
1:00 - 9:00	Public Skating

1. Seth is 9 years old. What time does his ice skating lesson begin?
2. For how long is the rink closed in the afternoon?
3. Natasha is 11 years old. She arrived at the rink at 9:55. How long is it until her lesson begins?
4. For how many hours a day is the rink open for lessons only?
5. What is the number of the group that is skating at 11:30?
6. For how many hours a day is the rink open for public skating?
7. Mrs. Kim teaches students ages 4 to 11. How long does she teach?

Riddle: What is a frog's favorite game?____________________

C	Y	A	R	O	E	S	Q	U	E	O	R	T	E
6	8	9:45	:55	12	2:15	:35	5	:25	9	3	4	:05	1 hour

Use the information on the chart to help you solve a problem. Cirlce the correct answers in the true or false columns. Each time the number of the problem appears at the bottom of the page, write the letter of the answer above it. When you are all done, decode the answer to the riddle.

Waterfalls of the World

Name	Location	Height (in meters)
Angel	Venezuela	979
Krimml	Austria	380
Ribbon	United States	491
Tugela	South Africa	914
Upper Yosemite	United States	436

TRUE	FALSE	
L	S	1. The tallest waterfall on the list is in Venezuela.
P	O	2. Two of the waterfalls on the list are located in South Africa.
S	T	3. The waterfall in Austria is shorter than Ribbon Falls.
N	M	4. Angel Falls is 65 meters taller than the waterfall in South Africa.
E	T	5. Angel Falls is more than twice the height of Ribbon Falls.
W	H	6. Upper Yosemite Falls is about 50 meters taller than Krimml Falls.
A	C	7. Tugela Falls is the second highest waterfall on the chart.
R	O	8. The waterfall in Austria is shorter than any other waterfall on the chart.
E	J	9. Ribbon Falls is the tallest U.S. waterfall on the chart.
D	A	10. Tugela Falls is more than twice the height of Upper Yosemite Falls.
H	U	11. Angel Falls is taller than both of the U.S. waterfalls together.
Y	E	12. There is over a 500 meter diference between the height of Krimml Falls and Angel Falls.

Riddle: How is a crossword puzzle like an argument?

__ __ __   __ __ __ __   __ __ __ __ __ __
2 4 9   6 2 8 10   7 1 6 7 12 3

__ __ __ __ __   __ __   __ __ __ __ __ __ __
1 9 7 10 3   5 2   7 4 2 5 11 9 8

PICTOGRAPHS

Use the information on the pictograph to help you solve a problem. Find the answer in the answer column. Each time the number of the problem appears at the bottom of the page, write the letter of the anwer above it. When you are all done, decode the answer to the riddle.

A pictograph uses pictures or drawings to compare information. The key at the bottom of a pictograph will tell you how many objects one picture stands for.

Answer Column:

T	**10**
S	**20**
Y	**30**
B	**4/18**
L	**sports**
N	**0**
E	**1/9**
H	**180**
A	**animals**
P	**history**

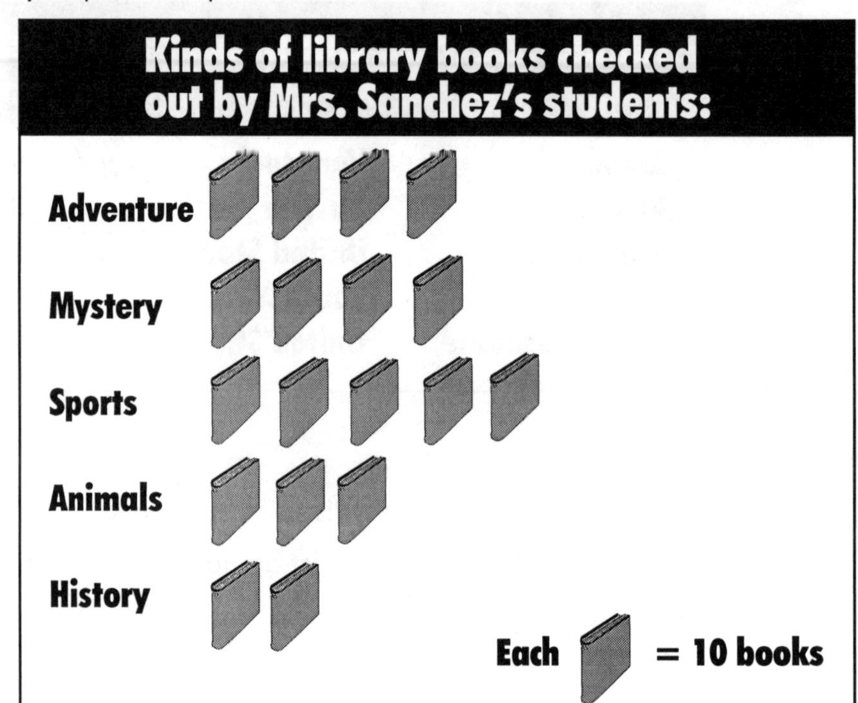

1. How many books in all were checked out?
2. How many more animal books were checked out than were history books?
3. In which category were a total of 50 books checked out?
4. In which category were the fewest books checked out?
5. How many more mystery books were checked out than history books?
6. In what category were a total of 30 books checked out?
7. What fraction of the total number of books checked out were mystery books?
8. How many more adventure books were checked out than mystery books?
9. What fraction of the total number of books checked out were history books?
10. If each student checked out 6 books, how many students are there in all?

Riddle: What do elephants have that no other animals have?

___ ___ ___ ___   ___ ___ ___ ___ ___ ___ ___ ___ ___

7 6 7 10   9 3 9 4 1 6 8 2 5

BAR GRAPHS

Find the answer to a question. Each time the answer appears at the bottom of the page, cross it out as well as the letter above it. When you are all done, unscramble the remaining letters to make a word that solves the riddle.

A bar graph uses bars of different lengths or heights to compare information. The bars on a bar graph can be drawn from side to side (called a horizontal bar graph) or from bottom to top (called a vertical bar graph). One box on the graph has the same value as every other box.

Lon asked his classmates to vote for the animal from his list that they would most want to have as a pet. This bar graph shows the results of his survey. Use the information on the graph to answer the questions.

1. How many of Lon's classmates chose a snake as their favorite pet?
2. How many classmates in all chose a pet that has legs?
3. How many more votes did Fish receive than did Rat?
4. How many votes in total were cast in Lon's survey?
5. This rodent received three votes.
6. How many votes does the longest bar on the graph stand for?
7. How many votes did Dog and Cat together receive?
8. What is another way of showing the number of votes Fish got?
9. What fraction of the total number of votes did Rat receive?
10. Which animal received 5 few votes than Cat?

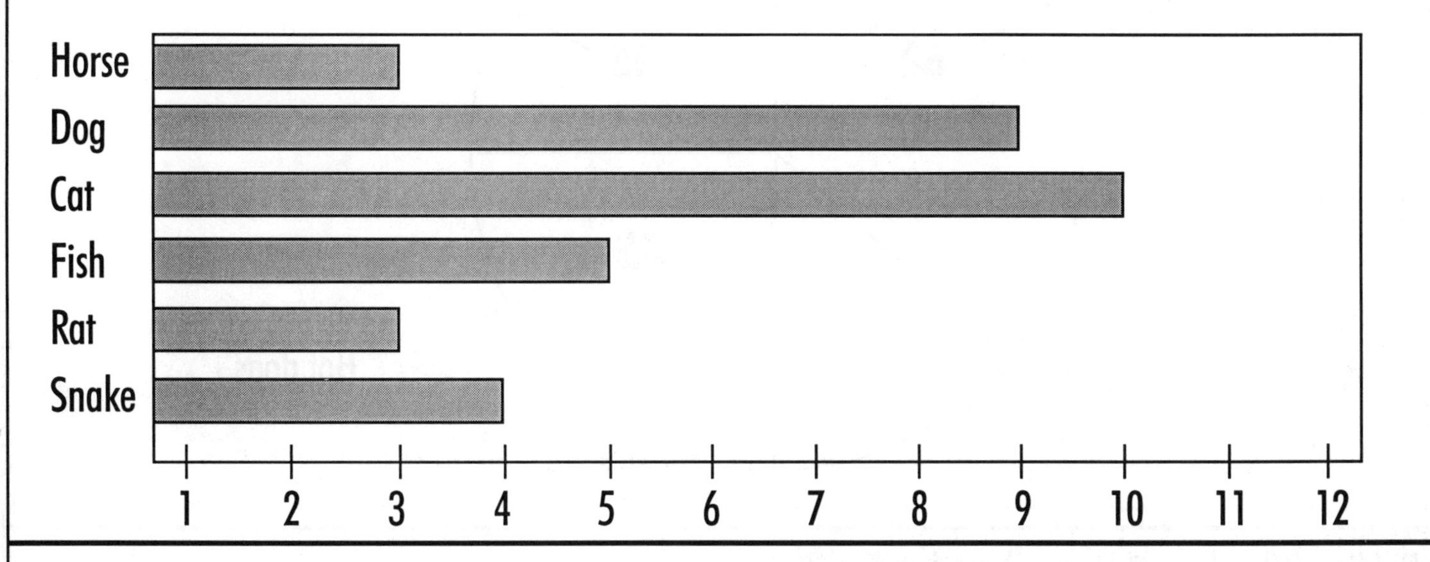

Question: What is the name of the greatest ocean liner in the world?

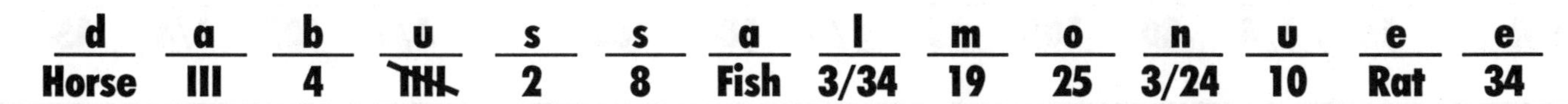

d	a	b	u	s	s	a	l	m	o	n	u	e	e
Horse	III	4	~~IIII~~	2	8	Fish	3/34	19	25	3/24	10	Rat	34

CIRCLE GRAPHS

Use the information on the circle graph to solve a problem. Each time the answer appears at the bottom of the page, write the letter of the problem above it. When you are all done, decode the answer to the riddle.

A circle graph shows how a total amount is divided into parts. It is easy to understand circle graphs if you think of them as pie or a pizza. The size of the slice tells how much of the whole is being represented.

- **M.** How many hot dogs were sold at the school carnival?
- **S.** How many more soft drinks were sold than were pretzels?
- **E.** How many cups in all were used to serve the soft drinks and fruit freezes?
- **A.** What fraction of the total sales were hot dogs?
- **T.** What fraction of the total sales were soft drinks and pretzels together?
- **B.** How many more soft drinks were sold than were hot dogs?
- **L.** How many snacks were sold in all at the school carnival?

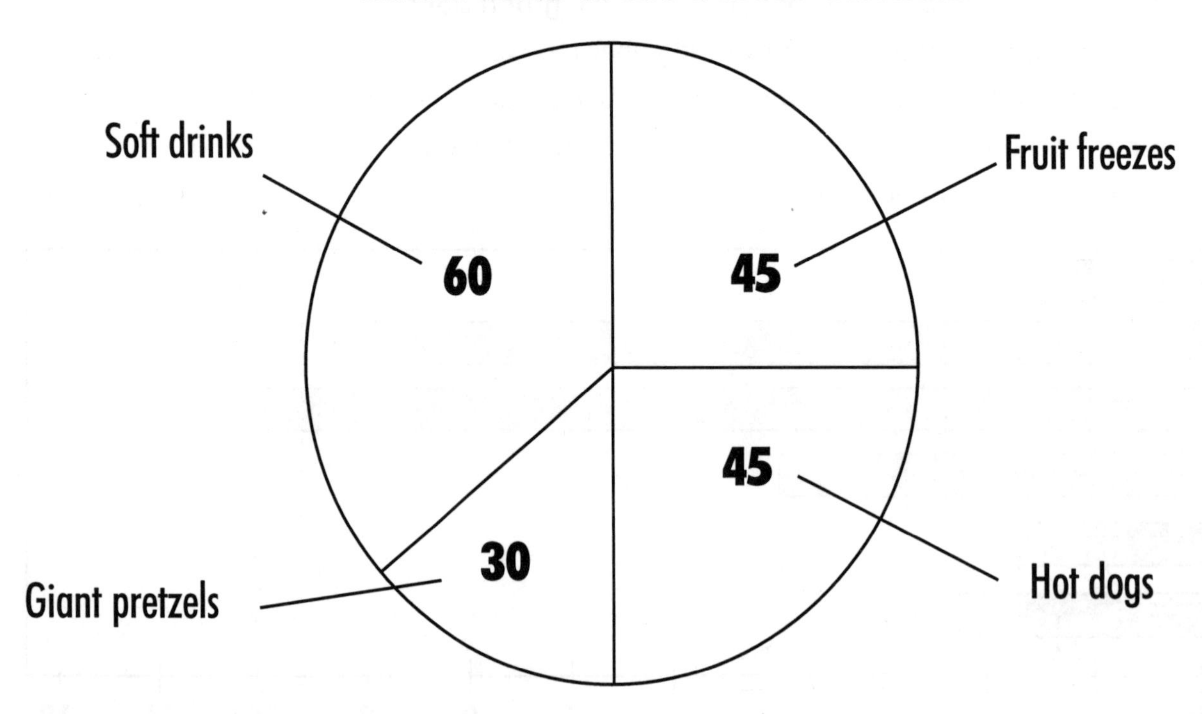

___
1/4

___ ___ ___ ___ ___ ___ ___ ___
15 1/4 30 105 15 1/4 180 180

___ ___ ___ ___
1/2 105 1/4 45

Answer Key

Please take time to go over the work your child has completed. Ask your child to explain what he/she has done. Praise both success and effort. If mistakes have been made, explain what the answer should have been and how to find it. Let your child know that mistakes are a part of learning. The time you spend with your child helps let him/her know you feel learning is important.

page 5

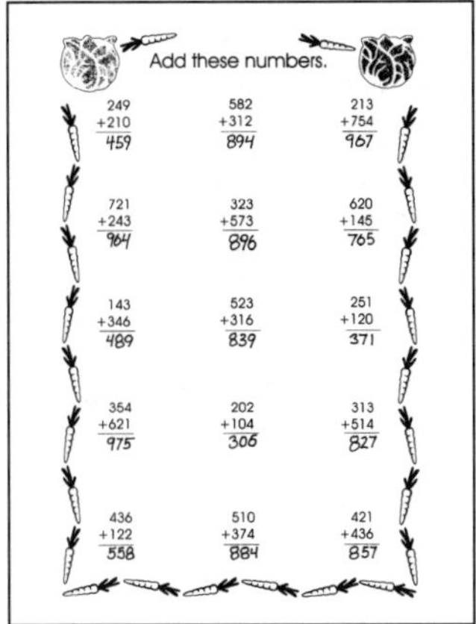

page 6

What has 18 legs, red spots and catches flies?

72 - a, 99 - h, 64 - s, 91 - b, 93 - i, 37 - t, 83 - e, 45 - L, 56 - w, 58 - m

49 +23

74 +17	27 +45	37 +27	55 +28	35 +56	56 +16	27 +18	19 +26
91	72	64	83	91	72	45	45
b	a	s	e	b	a	l	l

29 + 8	37 +46	35 +37	19 +39	28 +28	64 +29	19 +18	59 +40
37	83	72	58	56	93	37	99
t	e	a	m	w	i	t	h

26 +32	29 +54	60 +12	46 +18	15 +30	66 +17	15 +49
58	83	72	64	45	83	64
m	e	a	s	l	e	s

page 7

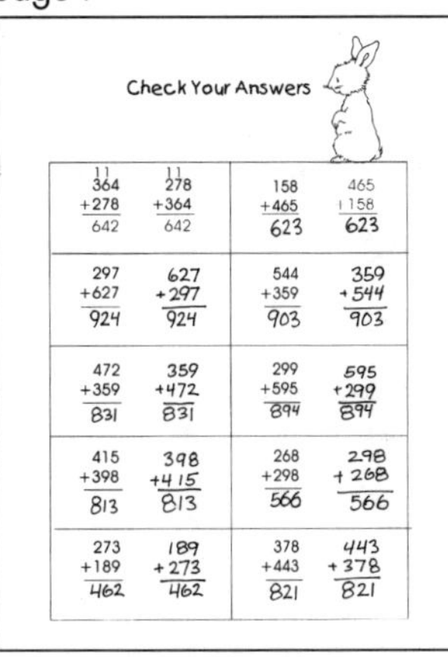

page 8

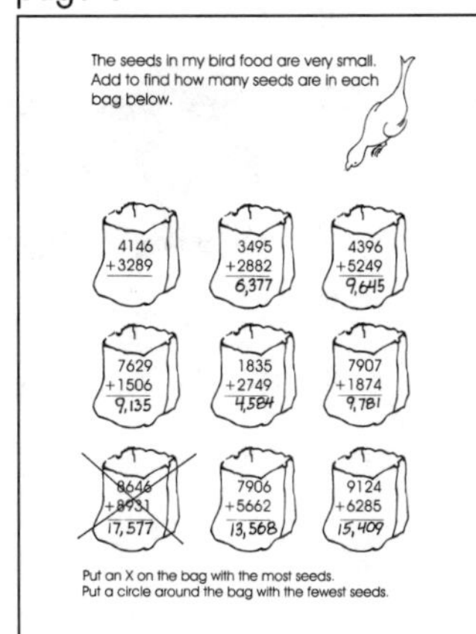

page 9

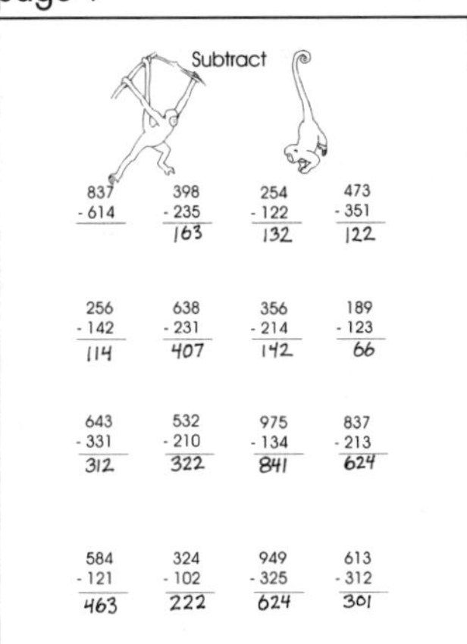

page 10

How do you know a gorilla was in your refrigerator?

18 - b, 44 - h, 78 - o, 89 - s, 56 - e, 15 - i, 63 - p, 29 - t, 36 - f, 47 - n, 27 - r, 62 - u

43 - 7	86 - 8	93 - 15	36 - 7
36	78	78	29
f	o	o	t

71 - 8	66 - 39	24 - 9	56 - 9	53 - 24	98 - 9
63	27	15	47	29	89
p	r	i	n	t	s

52 - 37	91 - 44		45 - 16	52 - 8	63 - 7
15	47		29	44	56
i	n		t	h	e

45 - 27	81 - 19	37 - 8	54 - 25	72 - 16	43 - 16
18	62	29	29	56	27
b	u	t	t	e	r

page 11

Help this little monkey answer some riddles.

47 - a, 78 - c, 179 - e, 227 - h, 245 - n, 338 - o, 387 - p, 395 - r, 427 - s, 577 - t

1. What has eyes but cannot see?

662 - 275	504 - 166	734 - 157	426 - 379	845 - 268	615 - 277
387	338	577	47	577	338
p	o	t	a	t	o

2. What has ears but cannot hear?

356 - 278	723 - 385	783 - 388	912 - 667
78	338	395	245
c	o	r	n

3. What has a tongue but cannot talk?

614 - 187	524 - 297	836 - 498	347 - 168
427	227	338	179
s	h	o	e

page 12

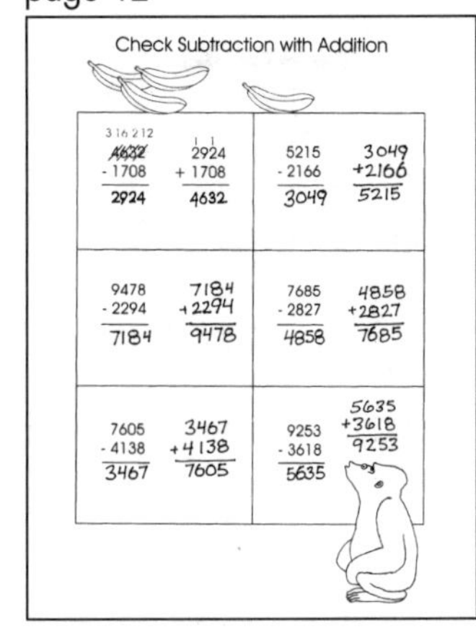

page 13

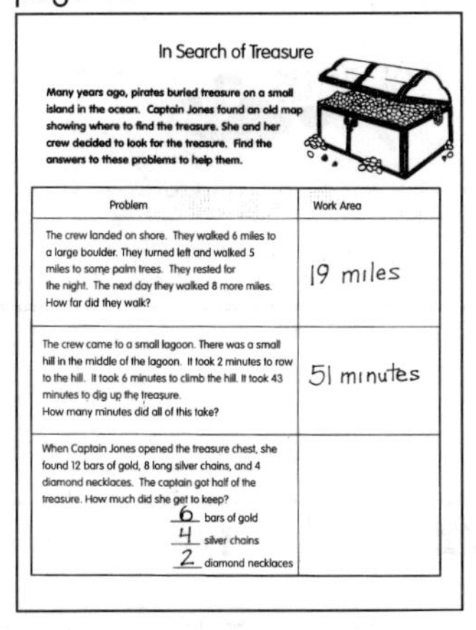

page 14

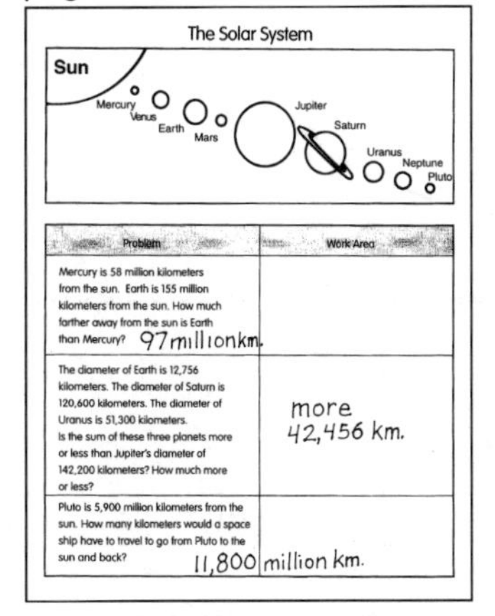

page 15

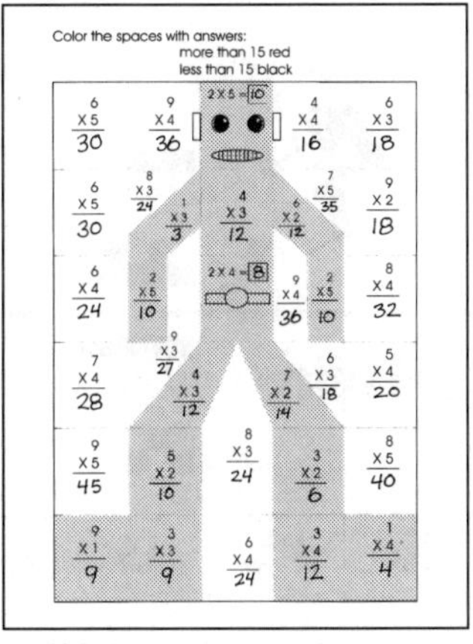

page 16

Multiply.

7 X 8	6 X 5	4 X 9	9 X 6	4 X 8	9 X 7
56	30	36	54	32	63

5 X 7	5 X 8	6 X 6	9 X 9	4 X 7	4 X 6
35	40	36	81	28	24

8 X 6	5 X 9	8 X 9	7 X 7	7 X 6	8 X 8
48	45	72	49	42	64

8 X 9 = 72	5 X 7 = 35	9 X 9 = 81
6 X 7 = 42	4 X 9 = 36	8 X 6 = 48
5 X 9 = 45	8 X 8 = 64	7 X 7 = 49
6 X 6 = 36	7 X 7 = 49	5 X 6 = 30
4 X 7 = 28	5 X 8 = 40	9 X 7 = 63

page 17

Two-Digit by One-Digit

Here's how to multiply a two-digit number by a one-digit number with regrouping (carrying):

13 × 4 = 52	12 × 8 = 96	14 × 7 = 98	19 × 5 = 95
18 × 9 = 162	16 × 6 = 96	14 × 3 = 42	17 × 9 = 153
24 × 2 = 48	23 × 3 = 69	24 × 2 = 48	30 × 5 = 150
42 × 4 = 168	31 × 7 = 217	24 × 4 = 96	62 × 2 = 124

page 18

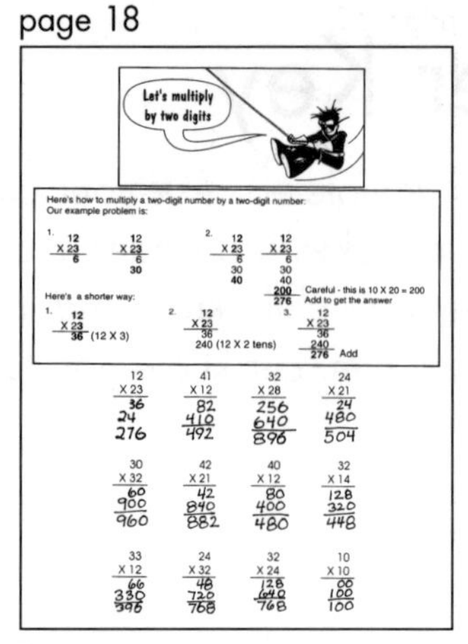

Here's how to multiply a two-digit number by a two-digit number:

12 × 23 = 276	41 × 12 = 492	32 × 28 = 896	24 × 21 = 504
30 × 32 = 960	42 × 21 = 882	40 × 12 = 480	32 × 14 = 448
33 × 12 = 396	24 × 32 = 768	32 × 24 = 768	10 × 10 = 100

page 19

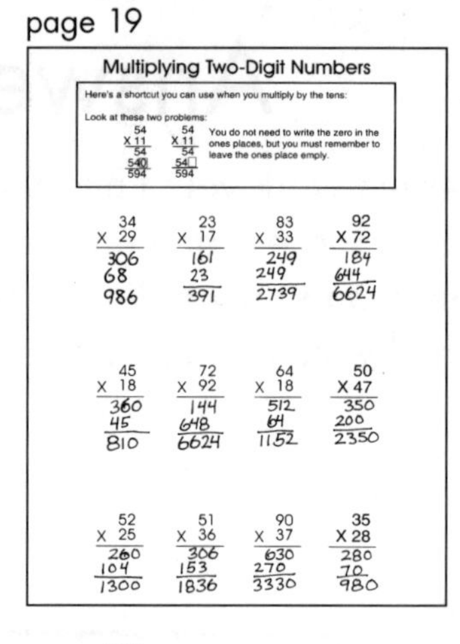
Multiplying Two-Digit Numbers

Here's a shortcut you can use when you multiply by the tens:

34 × 29 = 986	23 × 17 = 391	83 × 33 = 2739	92 × 72 = 6624
45 × 18 = 810	72 × 92 = 6624	64 × 18 = 1152	50 × 47 = 2350
52 × 25 = 1300	51 × 36 = 1836	90 × 37 = 3330	35 × 28 = 980

page 20

9 ÷ 3 = 3	2 ÷ 2 = 1	8 ÷ 4 = 2
45 ÷ 5 = 9	24 ÷ 3 = 8	25 ÷ 5 = 5
18 ÷ 6 = 3	56 ÷ 7 = 8	48 ÷ 8 = 6
27 ÷ 3 = 9	64 ÷ 8 = 8	9 ÷ 3 = 3
21 ÷ 7 = 3	18 ÷ 3 = 6	24 ÷ 8 = 3
12 ÷ 4 = 3	16 ÷ 8 = 2	54 ÷ 6 = 9
16 ÷ 8 = 2	40 ÷ 5 = 8	20 ÷ 4 = 5

page 21

page 22

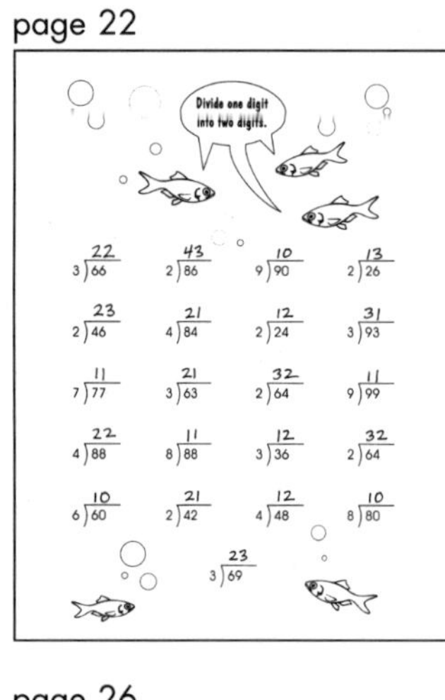

66 ÷ 3 = 22	86 ÷ 2 = 43	90 ÷ 9 = 10	26 ÷ 2 = 13
46 ÷ 2 = 23	84 ÷ 4 = 21	24 ÷ 2 = 12	93 ÷ 3 = 31
77 ÷ 7 = 11	63 ÷ 3 = 21	64 ÷ 2 = 32	99 ÷ 9 = 11
88 ÷ 4 = 22	88 ÷ 8 = 11	36 ÷ 3 = 12	64 ÷ 2 = 32
60 ÷ 6 = 10	42 ÷ 2 = 21	48 ÷ 4 = 12	80 ÷ 8 = 10
	69 ÷ 3 = 23		

page 23

Two-Step Divison

75 ÷ 3 = 25	78 ÷ 6 = 13	54 ÷ 2 = 27
90 ÷ 6 = 15	78 ÷ 3 = 26	70 ÷ 5 = 14
85 ÷ 5 = 17	72 ÷ 6 = 12	72 ÷ 4 = 18
91 ÷ 7 = 13	96 ÷ 4 = 24	84 ÷ 3 = 28

page 24

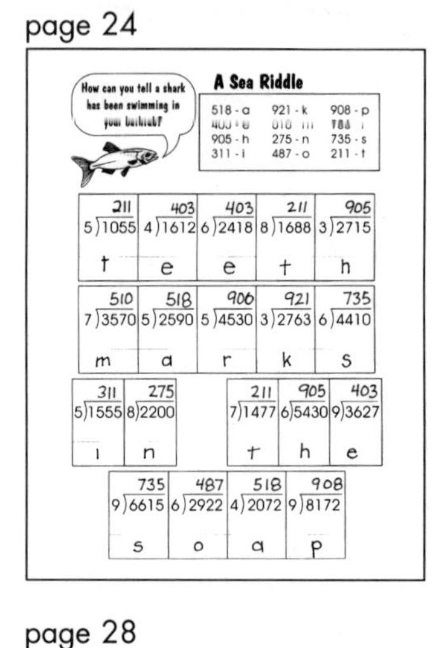
A Sea Riddle

1055 ÷ 5 = 211	1612 ÷ 4 = 403	2418 ÷ 6 = 403	1688 ÷ 8 = 211	2715 ÷ 3 = 905
t	e	e	t	h
3570 ÷ 7 = 510	2590 ÷ 5 = 518	4530 ÷ 5 = 906	2763 ÷ 3 = 921	4410 ÷ 6 = 735
m	a	r	k	s
1555 ÷ 5 = 311	2200 ÷ 8 = 275	1477 ÷ 7 = 211	5430 ÷ 6 = 905	3627 ÷ 9 = 403
i	n	t	h	e
6615 ÷ 9 = 735	2922 ÷ 6 = 487	2072 ÷ 4 = 518	8172 ÷ 9 = 908	
s	o	a	p	

page 25

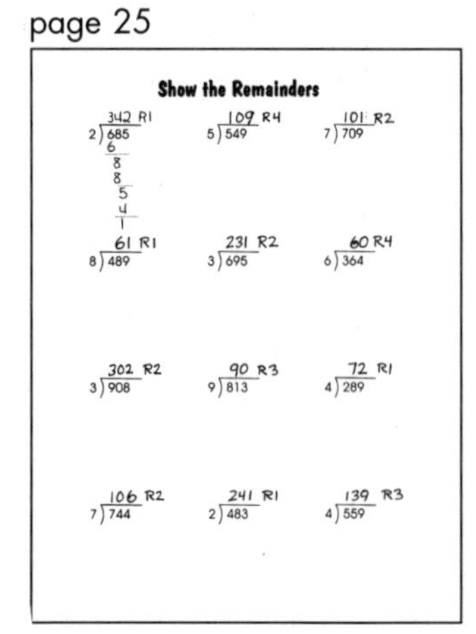
Show the Remainders

685 ÷ 2 = 342 R1	549 ÷ 5 = 109 R4	709 ÷ 7 = 101 R2
489 ÷ 8 = 61 R1	695 ÷ 3 = 231 R2	364 ÷ 6 = 60 R4
908 ÷ 3 = 302 R2	813 ÷ 9 = 90 R3	289 ÷ 4 = 72 R1
744 ÷ 7 = 106 R2	483 ÷ 2 = 241 R1	559 ÷ 4 = 139 R3

page 26

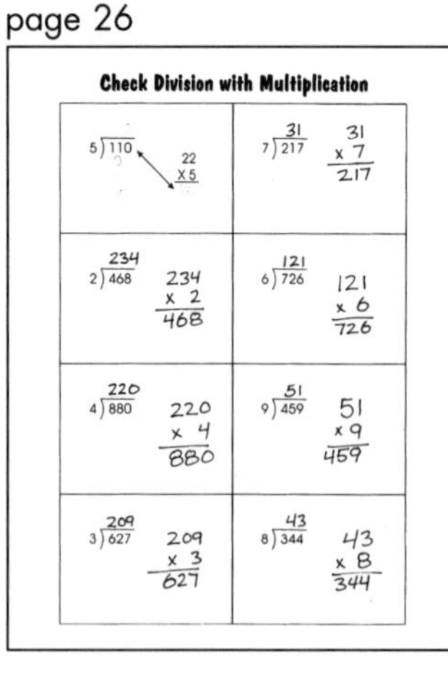
Check Division with Multiplication

110 ÷ 5 = 22; 22 × 5	217 ÷ 7 = 31; 31 × 7 = 217
468 ÷ 2 = 234; 234 × 2 = 468	726 ÷ 6 = 121; 121 × 6 = 726
880 ÷ 4 = 220; 220 × 4 = 880	459 ÷ 9 = 51; 51 × 9 = 459
627 ÷ 3 = 209; 209 × 3 = 627	344 ÷ 8 = 43; 43 × 8 = 344

page 27

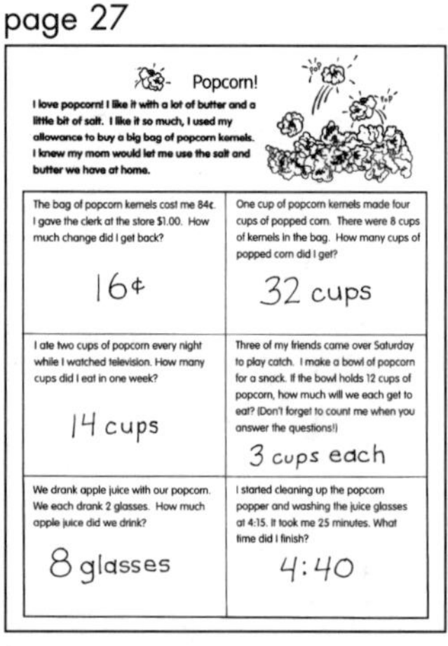
Popcorn!

I love popcorn! I like it with a lot of butter and a little bit of salt. I like it so much, I used my allowance to buy a big bag of popcorn kernels. I knew my mom would let me use the salt and butter we have at home.

The bag of popcorn kernels cost me 84¢. I gave the clerk at the store $1.00. How much change did I get back? 16¢	One cup of popcorn kernels made four cups of popped corn. There were 8 cups of kernels in the bag. How many cups of popped corn did I get? 32 cups
I ate two cups of popcorn every night while I watched television. How many cups did I eat in one week? 14 cups	Three of my friends came over Saturday to play catch. I make a bowl of popcorn for a snack. If the bowl holds 12 cups of popcorn, how much will we each get to eat? (Don't forget to count me when you answer the questions!) 3 cups each
We drank apple juice with our popcorn. We each drank 2 glasses. How much apple juice did we drink? 8 glasses	I started cleaning up the popcorn popper and washing the juice glasses at 4:15. It took me 25 minutes. What time did I finish? 4:40

page 28

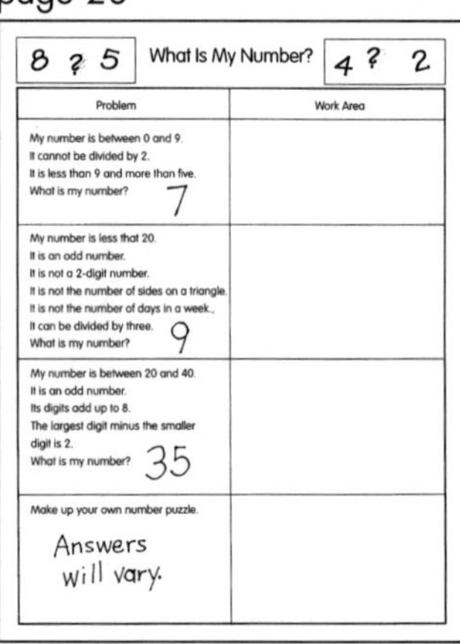
What Is My Number?

Problem	Work Area
My number is between 0 and 9. It cannot be divided by 2. It is less than 9 and more than five. What is my number? 7	
My number is less that 20. It is an odd number. It is not a 2-digit number. It is not the number of sides on a triangle. It is not the number of days in a week. It can be divided by three. What is my number? 9	
My number is between 20 and 40. It is an odd number. Its digits add up to 8. The largest digit minus the smaller digit is 2. What is my number? 35	
Make up your own number puzzle. Answers will vary.	

page 29

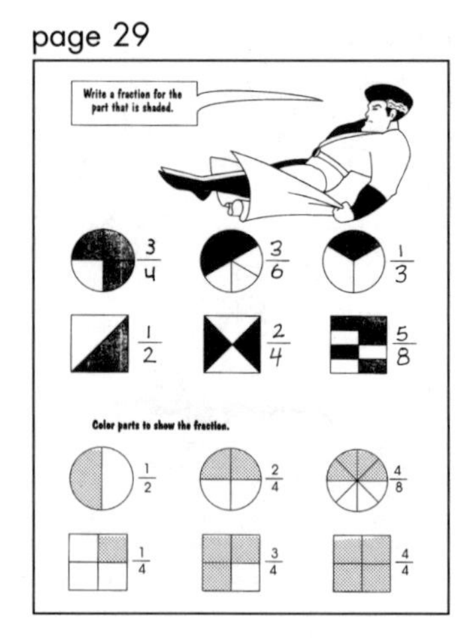

3/4, 3/6, 1/3

1/2, 2/4, 5/8

Color parts to show the fraction.

1/2, 2/4, 4/8

1/4, 3/4, 4/4

page 30

3/10, 7/10, 9/10, 5/10, 1/10, 4/10

Shade in parts to show the fraction.

1/2, 4/5

page 31

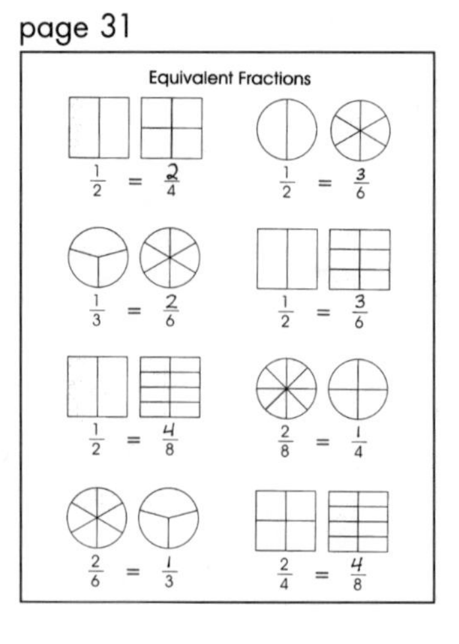
Equivalent Fractions

$\frac{1}{2} = \frac{2}{4}$ $\frac{1}{2} = \frac{3}{6}$

$\frac{1}{3} = \frac{2}{6}$ $\frac{1}{2} = \frac{3}{6}$

$\frac{1}{2} = \frac{4}{8}$ $\frac{2}{8} = \frac{1}{4}$

$\frac{2}{6} = \frac{1}{3}$ $\frac{2}{4} = \frac{4}{8}$

page 32

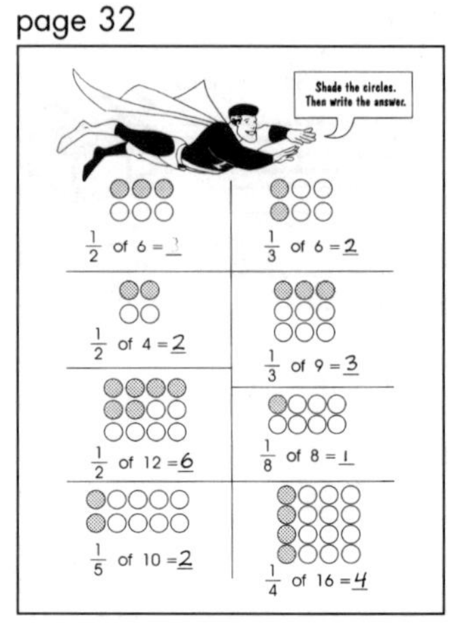

$\frac{1}{2}$ of 6 = 3	$\frac{1}{3}$ of 6 = 2
$\frac{1}{2}$ of 4 = 2	$\frac{1}{3}$ of 9 = 3
$\frac{1}{2}$ of 12 = 6	$\frac{1}{8}$ of 8 = 1
$\frac{1}{5}$ of 10 = 2	$\frac{1}{4}$ of 16 = 4

page 33

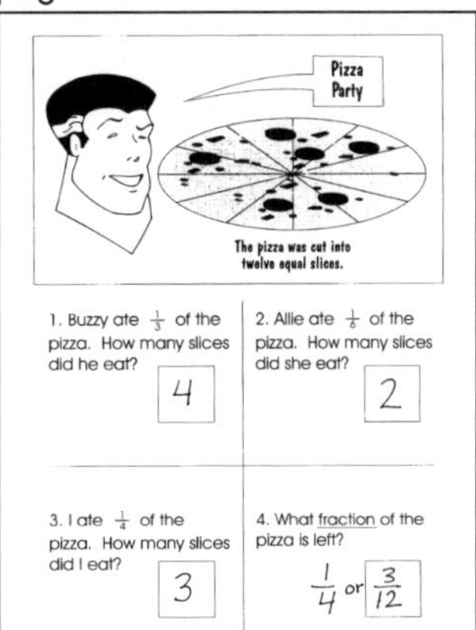

page 34

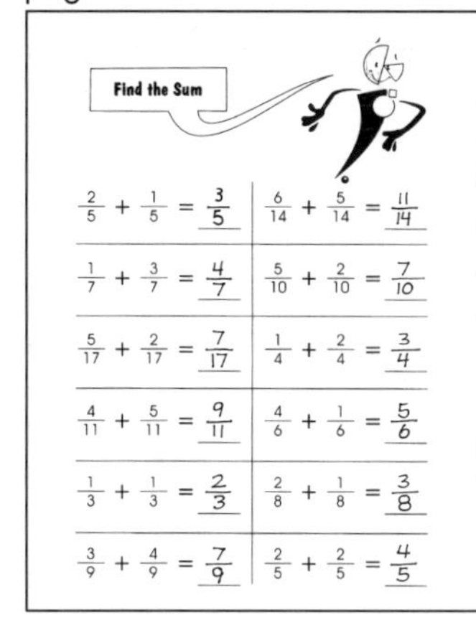

page 35

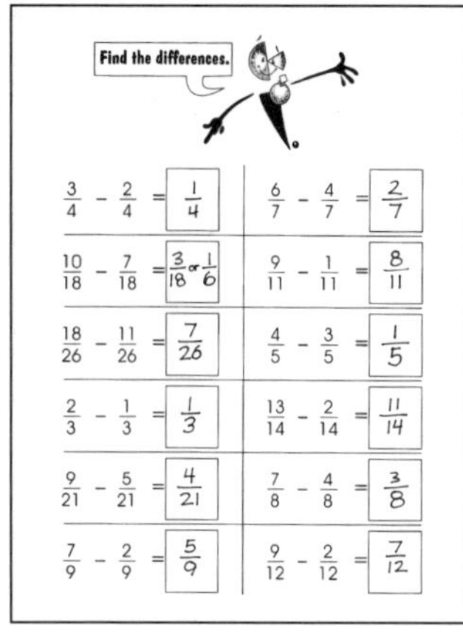

page 36

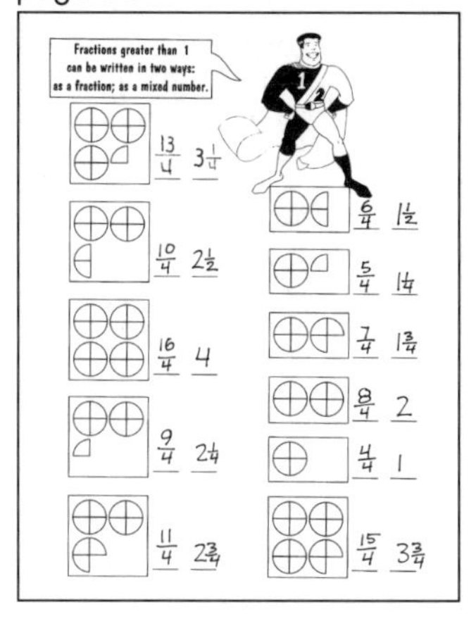

page 37

page 38

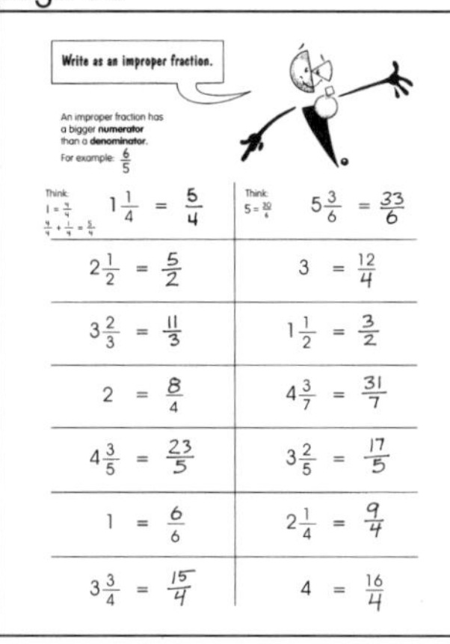

page 39

page 40

page 41

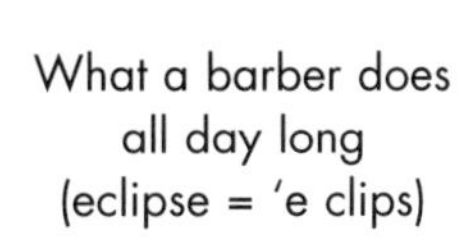

page 42

page 43

Davy Croakett
(Davy Crockett)

page 44

Robots have nerves
of steel.

page 45

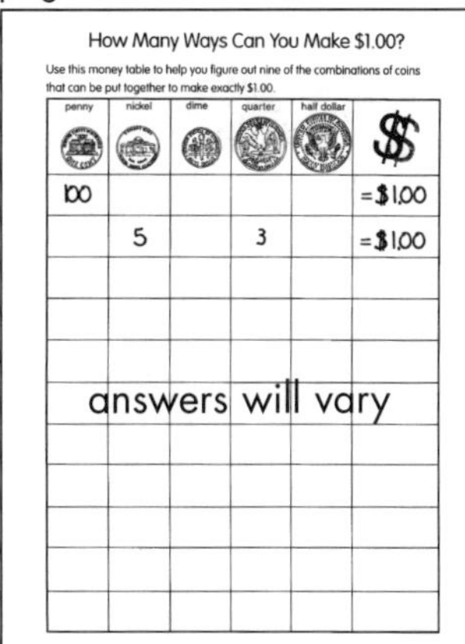

page 46

Don't count the minutes
at the table,
just the seconds

page 47

Headaches are all
in your mind

page 48

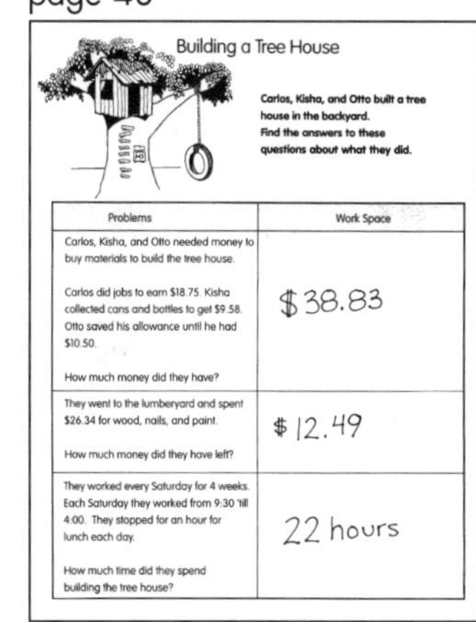

page 49

The letter "e" –
"every day" has
2 e's; "every week"
has 4 e's;
"a year" has 1 e

page 50

Both are twirly
(too early)

page 51

IOU
(I owe you)

page 52

North Poleish
(North Polish)

page 53

Come see come saw
(a take off on the
French phrase
"Comme cí, comme cá
meaning,
pretty good, OK")

page 55

Area =
720 square units

page 56

Croquet

page 57

One word always
leads to another

page 58

Baby elephants

page 59

Sand

page 60

A baseball team